Compact Textbooks in Mathematics

Compact Textbooks in Mathematics

This textbook series presents concise introductions to current topics in mathematics and mainly addresses advanced undergraduates and master students. The concept is to offer small books covering subject matter equivalent to 2- or 3-hour lectures or seminars which are also suitable for self-study. The books provide students and teachers with new perspectives and novel approaches. They may feature examples and exercises to illustrate key concepts and applications of the theoretical contents. The series also includes textbooks specifically speaking to the needs of students from other disciplines such as physics, computer science, engineering, life sciences, finance.

- **compact:** small books presenting the relevant knowledge
- **learning made easy:** examples and exercises illustrate the application of the contents
- **useful for lecturers:** each title can serve as basis and guideline for a semester course/lecture/seminar of 2–3 hours per week.

More information about this series at http://www.springer.com/series/11225

Piotr Sołtan

A Primer on Hilbert Space Operators

Piotr Sołtan
Faculty of Physics
University of Warsaw
Warsaw, Poland

ISSN 2296-4568 ISSN 2296-455X (electronic)
Compact Textbooks in Mathematics
ISBN 978-3-319-92060-3 ISBN 978-3-319-92061-0 (eBook)
https://doi.org/10.1007/978-3-319-92061-0

Library of Congress Control Number: 2018944130

Mathematics Subject Classification (2010): 47A05, 47A10, 47A60, 47B15, 47B25, 47B10, 47D03

This book is published under the imprint Birkhäuser, www.birkhauser-science.com by the registered company Springer Nature Switzerland AG
The registered company address is: Gewerbestrasse 11, 6330 Cham, Switzerland

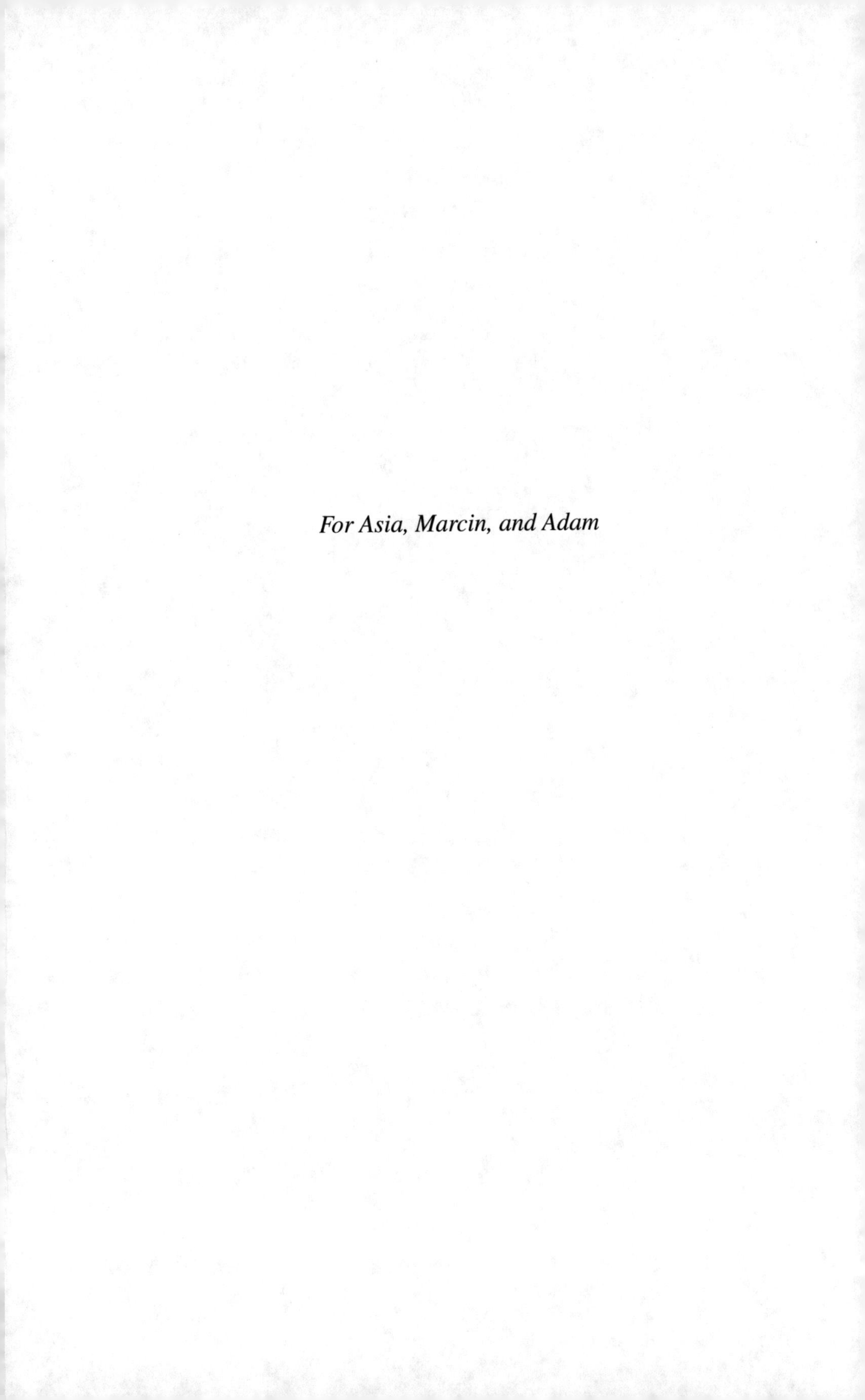

For Asia, Marcin, and Adam

Preface

Theory of operators on Hilbert spaces is one of the great achievements of functional analysis with applications in countless branches of mathematics and physics. It is also the basis of fascinating generalizations which include theory of C*-algebras, von Neumann algebras, non-commutative geometry, and many other domains of current research in mathematics. This book is based on extended lecture notes from a course held by the author at the Faculty of Physics of the University of Warsaw. Its main goals are:

- To provide a concise presentation of the rudiments of the theory of operators on Hilbert spaces with complete and direct proofs
- To prepare the reader for further study of operator theory itself as well as the theory of operator algebras (C*-algebras and von Neumann algebras)

The extent to which the first goal has been achieved can only be judged having defined the rudiments of the theory of operators on Hilbert spaces which tends to be a highly subjective matter. Some fragments of the theory have been purposefully left out in the hope of keeping the book's size moderate. More importantly, the reason for restricting the scope of the book to a bare minimum is the existence of a considerable number of unrivaled monographs and textbooks on the subject which cover a vastly larger range of material (e.g., [AkGl, Kat, ReSi$_1$, ReSi$_2$] and particularly [Mau]). However, studying these books might turn out to be quite challenging, and we hope that reading this modest volume might serve as a worthwhile preparatory exercise.

The material splits into two main parts. The first one is devoted to studying bounded operators, while the second deals with unbounded ones. In the latter part, we employ a novel and very useful tool introduced into world's mathematics by S.L. Woronowicz, namely, the so-called z-transform of a closed densely defined operator. Skipping ahead of the detailed introduction of the z-transform in ▶ Chap. 9, we can somewhat imprecisely describe it as a way to encode full information about a closed densely defined operator on a Hilbert space in a bounded operator on this space. The z-transform allows for simple and elegant proofs of many fundamental results of the theory.

Most of the excellent textbooks and monographs listed at the end of the book develop spectral theory of operators on Hilbert spaces based on several aspects of the theory of Banach algebras and C*-algebras. This approach requires the reader to work with rather sophisticated structures first and only then apply them to more elementary problems of operator theory. Our route is different: spectral theory of operators on Hilbert spaces is presented with minimal use of deeper results of Banach algebras. It is worth mentioning that, in fact, the theory of Banach algebras (in particular C*-algebras and von Neumann algebras) grew out of operator theory and can be viewed as its generalization *par excellence*. It is for this reason that we have chosen not to put too much emphasis on various aspects of the theory of Banach algebras.[1] It is the hope of the author that this approach will facilitate at least partial achievement of the second of the goals mentioned above.

Since the book aims to be a primer on theory of operators on Hilbert spaces, we decided not to include exercises nor many examples. Instead, we placed short notes at the end of each chapter containing references to textbooks and monographs containing a wealth of examples and exercises and, in some instances, information about possible further developments and generalizations of the subject of each chapter.

The lecture course on which the book is based was intended for students who have had previous experience with basic functional analysis including rudiments of Banach and Hilbert spaces. In particular, we assume the reader is familiar with:

- Basic linear algebra and calculus
- Elements of general topology including the concept of a locally compact topological space, a net and its convergence, and the Stone–Weierstrass theorem
- Elementary complex analysis including the notion of a holomorphic function, the Cauchy formula, and Liouville's theorem
- Theory of measure and integral including the dominated convergence theorem, product measures and Fubini's theorem, complex measures, Radon–Nikodym theorem, and Riesz–Markov–Kakutani representation theorem
- The concept of a Banach space and Hilbert space, L_p spaces, bounded operators on Banach spaces, and the operator norm
- The Riesz representation theorem (for functionals on Hilbert spaces), bounded sesquilinear forms and their relation to bounded operators on Hilbert spaces, and the notion of the adjoint operator of a bounded operator on a Hilbert space.

We will also make use of integrals of continuous Banach space-valued functions over compact intervals which are often discussed in courses of ordinary differential equations. A much more general theory of such integrals is presented e.g. in the monograph [Rud$_2$]. All of the above topics are covered in standard courses of complex analysis and measure theory, and textbooks such as e.g. [ReSi$_1$, Rud$_1$] discuss most of them. For the reader's convenience, we have gathered in the Appendix of the book some of the most important tools (inducing classical results such as the Banach-Steinhaus theorem and the closed graph theorem) with complete proofs.

[1] One exception from this rule comes in ▶ Chap. 7 in which we apply several results of C*-algebra theory. The results in question have been gathered in Appendix A.5.2.

All vector spaces considered in this book will be over the field of complex numbers. We will employ the conventions of the physics literature according to which scalar products will be linear in the second variable and anti-linear in the first. We will also use the "ket" and "bra" notation which we will now briefly discuss. Let $\mathcal{H}$ be a Hilbert space. Then any vector $\psi \in \mathcal{H}$ defines a unique linear operator $\mathbb{C} \to \mathcal{H}$ mapping $1 \in \mathbb{C}$ to ψ. We denote this linear map by the symbol $|\psi\rangle$.

Now consider on $\mathbb{C}$ the standard Hilbert space structure (one for which $\{1\}$ is an orthonormal basis). Then the adjoint $|\psi\rangle^*$ of $|\psi\rangle$ is a bounded linear functional on $\mathcal{H}$ mapping any vector ϕ onto the number $\langle\psi|\phi\rangle$. This map is denoted by $\langle\psi|$. In particular, the composition $\langle\psi_1| \circ |\psi_2\rangle$ is a linear map $\mathbb{C} \to \mathbb{C}$ given by multiplication by the scalar $\langle\psi_1|\psi_2\rangle$, while the composition $|\psi_2\rangle \circ \langle\psi_1|$ (customarily written as $|\psi_2\rangle\langle\psi_1|$) is a map $\mathcal{H} \to \mathcal{H}$ taking $\varphi \in \mathcal{H}$ to $\langle\psi_1|\varphi\rangle\,\psi_2$.

One of the fundamental formulas from the theory of vector spaces endowed with a sesquilinear form is the *polarization formula*. It has various, sometimes highly sophisticated, formulations one of which will be especially useful to us: let F be a sesquilinear form on a vector space $\mathcal{H}$ (linear with respect to the second argument). Then

$$F(\xi,\eta) = \tfrac{1}{4}\sum_{k=0}^{3} \mathrm{i}^k F(\eta + \mathrm{i}^k\xi, \eta + \mathrm{i}^k\xi), \qquad \xi,\eta \in \mathcal{H}.$$

The material covered in this book is for the most part linearly ordered, i.e. each chapter uses results established in preceding ones. The notable exception to this rule are ▶ Chap. 6 devoted to the theory of the trace and ▶ Chap. 7 dealing with functional calculus for families of self-adjoint operators and for normal operators. The results of these chapters are not used in the remaining parts of the book. Also the results of ▶ Chap. 10 on spectral theory of unbounded operators are not used until ▶ Chap. 12.

As we already mentioned, the Appendix contains additional material needed in various places in the book arranged into five sections.

I wish to thank my teacher professor S.L. Woronowicz who introduced me to the theory of operators on Hilbert spaces and has over the years shared with me his knowledge of the subject. I also thank colleagues and students from the Faculty of Physics and from the Department of Mathematics of the University of Warsaw for their support and helpful remarks during the writing of this book.

Piotr Mikołaj Sołtan

Warsaw
July 2018

Contents

Part I Bounded Operators

Part II Unbounded Operators

Part I
Bounded Operators

Spectrum of an Operator

P. Sołtan, *A Primer on Hilbert Space Operators*, Compact Textbooks in Mathematics,
https://doi.org/10.1007/978-3-319-92061-0_1

1.1 C*-Algebra of Operators on a Hilbert Space

Let $\mathcal{H}$ be a Hilbert space. The Banach space $\mathsf{B}(\mathcal{H})$ of bounded operators on $\mathcal{H}$ is an algebra over $\mathbb{C}$ with multiplication defined as composition of operators and the identity operator $\mathbb{1} : \mathcal{H} \to \mathcal{H}$ is the unit (neutral element of multiplication) of this algebra. The operation of passing to the adjoint operator

$$\mathsf{B}(\mathcal{H}) \ni x \longmapsto x^* \in \mathsf{B}(\mathcal{H})$$

is an anti-linear, anti-multiplicative *involution* (for any $x \in \mathsf{B}(\mathcal{H})$ we have $x^{**} = x$). Moreover, the operator norm is compatible with algebra structure in the sense that

$$\|xy\| \leq \|x\|\|y\|, \qquad x, y \in \mathsf{B}(\mathcal{H}).$$

In particular $\mathsf{B}(\mathcal{H})$ is a *Banach algebra.*

Proposition 1.1 *For $x \in \mathsf{B}(\mathcal{H})$ we have*
(1) $\|x\| = \|x^*\|$,
(2) $\|x^*x\| = \|x\|^2$.

Proof
Both equalities are obvious for $x = 0$. Therefore let us assume that $\|x\| > 0$. Then clearly

$$\|x^*x\| \leq \|x^*\|\|x\|.$$

Furthermore, the computation

$$\begin{aligned}\|x^*x\| = \sup_{\|\xi\|=1} \|x^*x\xi\| &= \sup_{\|\xi\|=1} \sup_{\|\eta\|=1} \left|\langle \eta | x^*x\xi \rangle\right| \\ &\geq \sup_{\|\xi\|=1} \left|\langle \xi | x^*x\xi \rangle\right| = \sup_{\|\xi\|=1} \|x\xi\|^2 = \|x\|^2.\end{aligned}$$

yields

$$\|x\|^2 \leq \|x^*x\| \leq \|x^*\|\|x\|. \tag{1.1}$$

Dividing both sides by $\|x\|$ we obtain

$$\|x\| \leq \|x^*\|,$$

which by symmetry shows that $\|x^*\| = \|x\|$. Substituting this into (1.1) gives $\|x^*x\| = \|x\|^2$.

□

Remark 1.2 Let us note that the proof of Proposition 1.1 can easily be extended to the case when x is an operator between different Hilbert spaces. Thus if $\mathcal{H}$ and $\mathcal{K}$ are Hilbert spaces then for any $x \in \mathsf{B}(\mathcal{H}, \mathcal{K})$ we have $\|x^*x\| = \|x\|^2$.

Proposition 1.1(1) says that the involution $x \mapsto x^*$ on $\mathsf{B}(\mathcal{H})$ is isometric. A Banach algebra together with an isometric, anti-linear and anti-multiplicative involution is called a *Banach ∗-algebra*, while a Banach ∗-algebra whose involution additionally possesses property (2) from Proposition 1.1 is called a *C*-algebra*. Thus $\mathsf{B}(\mathcal{H})$ is a C*-algebra.

Moreover any norm-closed ∗-subalgebra[1] $\mathsf{A} \subset \mathsf{B}(\mathcal{H})$ is also a C*-algebra. For any subset $S \subset \mathsf{B}(\mathcal{H})$ there exists the smallest C*-algebra $\mathsf{A} \subset \mathsf{B}(\mathcal{H})$ containing S and it is called the C*-algebra *generated* by S. We denote it by the symbol $\mathrm{C}^*(S)$. It is easy to show that $\mathrm{C}^*(S)$ is the closure of the set of linear combinations of all products of elements of the sets S and $S^* = \{s^* \mid s \in S\}$. For $x \in \mathsf{B}(\mathcal{H})$ we will write $\mathrm{C}^*(x)$ and $\mathrm{C}^*(x, \mathbb{1})$ instead of $\mathrm{C}^*(\{x\})$ and $\mathrm{C}^*(\{x, \mathbb{1}\})$.

Another example of a C*-algebra is the space $\mathrm{C}(X)$ of continuous functions on a compact space X with uniform norm $\|\cdot\|_\infty$, pointwise addition and multiplication and involution given by $f \mapsto \overline{f}$. A seemingly different example would be the space $\mathrm{C}_b(Y)$ of bounded continuous functions on a locally compact space Y with norm $\|\cdot\|_\infty$ and pointwise algebraic operations. However this example is not really different from the previous one, as in fact $\mathrm{C}_b(Y)$ is naturally isomorphic to $\mathrm{C}(\beta Y)$, where βY is the Stone-Čech compactification of Y.[2]

1.2 Spectrum and Spectral Radius

Let $x \in \mathsf{B}(\mathcal{H})$. Recall that x is *invertible* if there exists an operator $y \in \mathsf{B}(\mathcal{H})$ such that $xy = yx = \mathbb{1}$. The *resolvent set* of x is

$$\rho(x) = \bigl\{\lambda \in \mathbb{C} \bigm| \text{the operator } \lambda\mathbb{1} - x \text{ is invertible}\bigr\},$$

and its complement $\sigma(x) = \mathbb{C} \setminus \rho(x)$ is called the *spectrum* of x.

[1] Vector subspace closed under composition of operators and passing to the adjoint operator.
[2] Cf. [Eng, Corollary 3.6.3].

It is known that the set of invertible operators is open (see below for an argument proving this) and since $\rho(x)$ is the pre-image of this set under the continuous map

$$\mathbb{C} \ni \lambda \longmapsto \lambda\mathbb{1} - x \in \mathsf{B}(\mathcal{H}),$$

we see that the resolvent set is open. Moreover, if $\lambda_0 \in \rho(x)$ and $\lambda \in \mathbb{C}$ satisfies

$$|\lambda - \lambda_0| < \tfrac{1}{\|(\lambda_0\mathbb{1}-x)^{-1}\|},$$

then $\lambda \in \rho(x)$ and it is easy to see that

$$(\lambda\mathbb{1} - x)^{-1} = \sum_{n=0}^{\infty}(\lambda_0 - \lambda)^n(\lambda_0\mathbb{1} - x)^{-n-1}.$$

In particular the mapping $\rho(x) \ni \lambda \mapsto (\lambda\mathbb{1} - x)^{-1} \in \mathsf{B}(\mathcal{H})$ called the *resolvent* of x is holomorphic.

Remark 1.3 For any $\lambda, \mu \in \rho(x)$ we have

$$(\lambda\mathbb{1} - x)^{-1} - (\mu\mathbb{1} - x)^{-1} = (\mu - \lambda)(\lambda\mathbb{1} - x)^{-1}(\mu\mathbb{1} - x)^{-1}. \tag{1.2}$$

(in particular the values of the resolvent of x at different points of $\rho(x)$ commute). Indeed: the formula is obvious for $\lambda = \mu$ and for $\lambda \neq \mu$ we easily check that

$$\begin{aligned}
&\tfrac{1}{\mu-\lambda}\big((\lambda\mathbb{1} - x)^{-1} - (\mu\mathbb{1} - x)^{-1}\big)(\mu\mathbb{1} - x)(\lambda\mathbb{1} - x)\\
&\qquad= \tfrac{1}{\mu-\lambda}\big((\lambda\mathbb{1} - x)^{-1}(\mu\mathbb{1} - x) - \mathbb{1}\big)(\lambda\mathbb{1} - x)\\
&\qquad= \tfrac{1}{\mu-\lambda}\big((\lambda\mathbb{1} - x)^{-1}\big((\mu - \lambda)\mathbb{1} + (\lambda\mathbb{1} - x)\big) - \mathbb{1}\big)(\lambda\mathbb{1} - x)\\
&\qquad= \tfrac{1}{\mu-\lambda}\big((\mu - \lambda)(\lambda\mathbb{1} - x)^{-1} + \mathbb{1} - \mathbb{1}\big)(\lambda\mathbb{1} - x) = \mathbb{1}
\end{aligned}$$

and

$$\begin{aligned}
&(\mu\mathbb{1} - x)(\lambda\mathbb{1} - x)\tfrac{1}{\mu-\lambda}\big((\lambda\mathbb{1} - x)^{-1} - (\mu\mathbb{1} - x)^{-1}\big)\\
&\qquad= \tfrac{1}{\mu-\lambda}(\mu\mathbb{1} - x)\big(\mathbb{1} - (\lambda\mathbb{1} - x)(\mu\mathbb{1} - x)^{-1}\big)\\
&\qquad= \tfrac{1}{\mu-\lambda}(\mu\mathbb{1} - x)\big(\mathbb{1} - \big((\lambda - \mu)\mathbb{1} + (\mu\mathbb{1} - x)\big)(\mu\mathbb{1} - x)^{-1}\big)\\
&\qquad= \tfrac{1}{\mu-\lambda}(\mu\mathbb{1} - x)\big(\mathbb{1} - (\lambda - \mu)(\mu\mathbb{1} - x)^{-1} + \mathbb{1}\big) = \mathbb{1}.
\end{aligned}$$

It follows that $(\mu\mathbb{1} - x)(\lambda\mathbb{1} - x)$ is invertible and its inverse is the operator $\tfrac{1}{\mu-\lambda}\big((\lambda\mathbb{1} - x)^{-1} - (\mu\mathbb{1} - x)^{-1}\big)$. Formula (1.2) is known as the *resolvent identity* or the *resolvent formula*.

Proposition 1.4 *Let* $x, y \in \mathsf{B}(\mathcal{H})$. *Then*

$$\sigma(xy) \cup \{0\} = \sigma(yx) \cup \{0\}. \tag{1.3}$$

Proof

Let $\lambda \in \rho(yx) \setminus \{0\}$, so that the operator $\lambda\mathbb{1} - yx$ is invertible. We have

$$\begin{aligned}
(\lambda\mathbb{1} - xy)\Big(\tfrac{1}{\lambda}\big(\mathbb{1} + x(\lambda\mathbb{1} - yx)^{-1}y\big)\Big) \\
&= \tfrac{1}{\lambda}\big(\lambda\mathbb{1} - xy + (\lambda\mathbb{1} - xy)x(\lambda\mathbb{1} - yx)^{-1}y\big) \\
&= \tfrac{1}{\lambda}\big(\lambda\mathbb{1} - xy + x(\lambda\mathbb{1} - yx)(\lambda\mathbb{1} - yx)^{-1}y\big) \\
&= \tfrac{1}{\lambda}(\lambda\mathbb{1} - xy + xy) = \mathbb{1}
\end{aligned}$$

and

$$\begin{aligned}
\Big(\tfrac{1}{\lambda}\big(\mathbb{1} + x(\lambda\mathbb{1} - yx)^{-1}y\big)\Big)(\lambda\mathbb{1} - xy) \\
&= \tfrac{1}{\lambda}\big(\lambda\mathbb{1} - xy + x(\lambda\mathbb{1} - yx)^{-1}y(\lambda\mathbb{1} - xy)\big) \\
&= \tfrac{1}{\lambda}\big(\lambda\mathbb{1} - xy + x(\lambda\mathbb{1} - yx)^{-1}(\lambda\mathbb{1} - yx)y\big) \\
&= \tfrac{1}{\lambda}(\lambda\mathbb{1} - xy + xy) = \mathbb{1}
\end{aligned}$$

which means that $\lambda\mathbb{1} - xy$ is invertible with inverse $\frac{1}{\lambda}\big(\mathbb{1} + x(\lambda\mathbb{1} - yx)^{-1}y\big)$. In other words $\lambda \in \rho(xy) \setminus \{0\}$ and this shows that $\rho(yx) \setminus \{0\} \subset \rho(xy) \setminus \{0\}$. By symmetry, also $\rho(xy) \setminus \{0\} \subset \rho(yx) \setminus \{0\}$ and hence $\rho(xy) \setminus \{0\} = \rho(yx) \setminus \{0\}$. In other words $\sigma(xy) \cup \{0\} = \sigma(yx) \cup \{0\}$. □

Remark 1.5 By taking set difference of both sides of (1.3) with $\{0\}$ we get another useful expression of the same phenomenon:

$$\sigma(xy) \setminus \{0\} = \sigma(yx) \setminus \{0\}.$$

Proposition 1.6 *For any* $x \in \mathsf{B}(\mathcal{H})$ *we have* $\big\{\lambda \in \mathbb{C} \,\big|\, |\lambda| > \|x\|\big\} \subset \rho(x)$.

Proof

If $|\lambda| > \|x\|$ then the series

$$\sum_{n=0}^{\infty} \lambda^{-n-1} x^n = \tfrac{1}{\lambda} \sum_{n=0}^{\infty} \lambda^{-n} x^n$$

converges in $\mathsf{B}(\mathcal{H})$, because $\|\lambda^{-n} x^n\| \leq \Big(\frac{\|x\|}{|\lambda|}\Big)^n$. One easily checks that its sum is the inverse of $\lambda\mathbb{1} - x$. □

Rephrasing Proposition 1.6, for any $x \in \mathsf{B}(\mathcal{H})$ we have

$$\sigma(x) = \mathbb{C} \setminus \rho(x) \subset \mathbb{C} \setminus \{\lambda \in \mathbb{C} \,|\, |\lambda| > \|x\|\} = \{\lambda \in \mathbb{C} \,|\, |\lambda| \leq \|x\|\},$$

and therefore the spectrum of x is a closed and bounded (hence compact) subset of $\mathbb{C}$.

Theorem 1.7
For any $x \in \mathsf{B}(\mathcal{H})$ we have
(1) *$\sigma(x)$ is non-empty,*
(2) *the sequence $\big(\|x^n\|^{\frac{1}{n}}\big)_{n\in\mathbb{N}}$ is convergent and*

$$\lim_{n\to\infty} \|x^n\|^{\frac{1}{n}} = \sup\{|\lambda| \,|\, \lambda \in \sigma(x)\}.$$

Proof
Define

$$\alpha(x) = \inf_{n\in\mathbb{N}} \|x^n\|^{\frac{1}{n}}, \qquad x \in \mathsf{B}(\mathcal{H}).$$

Take any $\varepsilon > 0$. Then there is a natural number $n_\varepsilon \in \mathbb{N}$ such that $\|x^{n_\varepsilon}\|^{\frac{1}{n_\varepsilon}} \leq \alpha(x) + \varepsilon$ or, in other words,

$$\|x^{n_\varepsilon}\| \leq \big(\alpha(x) + \varepsilon\big)^{n_\varepsilon}.$$

Take now any $n \in \mathbb{N}$ and divide it by n_ε with remainder, i.e. find $q, r \in \mathbb{Z}_+$ such that

$$n = qn_\varepsilon + r$$

and $r < n_\varepsilon$. Then we have

$$\|x^n\| = \|x^{qn_\varepsilon} x^r\| \leq \|x^{n_\varepsilon}\|^q \|x\|^r \leq \big(\alpha(x) + \varepsilon\big)^{qn_\varepsilon} \|x\|^r = \big(\alpha(x) + \varepsilon\big)^{n-r} \|x\|^r.$$

Therefore

$$\|x^n\|^{\frac{1}{n}} \leq \big(\alpha(x) + \varepsilon\big)^{1-\frac{r}{n}} \|x\|^{\frac{r}{n}}$$

which shows that

$$\alpha(x) \leq \liminf_{n\to\infty} \|x^n\|^{\frac{1}{n}} \leq \limsup_{n\to\infty} \|x^n\|^{\frac{1}{n}} \leq \alpha(x) + \varepsilon.$$

As ε is arbitrary, we find that the sequence $\big(\|x^n\|^{\frac{1}{n}}\big)_{n\in\mathbb{N}}$ is convergent (with limit $\alpha(x)$).

In particular, if $x, y \in \mathsf{B}(\mathcal{H})$ commute then

$$\alpha(xy) = \lim_{n\to\infty} \left\|(xy)^n\right\|^{\frac{1}{n}} = \lim_{n\to\infty} \|x^n y^n\|^{\frac{1}{n}}$$
$$\leq \lim_{n\to\infty} \|x^n\|^{\frac{1}{n}} \|y^n\|^{\frac{1}{n}} = \lim_{n\to\infty} \|x^n\|^{\frac{1}{n}} \lim_{n\to\infty} \|x^n\|^{\frac{1}{n}} = \alpha(x)\alpha(y).$$

Now if $|\lambda| > \alpha(x)$ then the series $\sum_{n=0}^{\infty} \frac{\|x^n\|}{|\lambda|^n}$ converges, and therefore so does

$$\sum_{n=0}^{\infty} \lambda^{-n} x^n.$$

We easily check that its sum is the inverse of $\mathbb{1} - \frac{x}{\lambda}$. It follows that $\lambda\mathbb{1} - x$ is invertible and we obtain

$$\alpha(x) \geq \sup\left\{|\lambda| \,\middle|\, \lambda \in \sigma(x)\right\}. \tag{1.4}$$

For the proof of the reverse inequality and non-emptiness of the spectrum we need to consider two cases.

Case 1 $\alpha(x) = 0$. In this case x is not invertible (and so $0 \in \sigma(x)$). Indeed: if x were invertible we would have

$$1 = \alpha(\mathbb{1}) = \alpha(xx^{-1}) \leq \alpha(x)\alpha(x^{-1}) = 0.$$

Case 2 $\alpha(x) > 0$. Let us assume that $\alpha(x) > \sup\left\{|\lambda| \,\middle|\, \lambda \in \sigma(x)\right\}$. Since $\sigma(x)$ is a compact subset of $\mathbb{C}$, there exists $r \in \left]0, \alpha(x)\right[$ with the property that

$$\sigma(x) \subset \left\{\lambda \in \mathbb{C} \,\middle|\, |\lambda| \leq r\right\}.$$

Thus the set $D = \left\{\lambda \in \mathbb{C} \,\middle|\, |\lambda| > r\right\}$ is contained in $\rho(x)$. For any continuous functional φ on $\mathsf{B}(\mathcal{H})$ the function

$$D \ni \lambda \longmapsto \varphi\big((\lambda\mathbb{1} - x)^{-1}\big) \in \mathbb{C}$$

is holomorphic. Moreover, for $|\lambda| > \alpha(x)$ we have

$$\varphi\big((\lambda\mathbb{1} - x)^{-1}\big) = \sum_{n=0}^{\infty} \lambda^{-n-1} \varphi(x^n).$$

This function vanishes at infinity,[3] so

$$f(\mu) = \begin{cases} 0, & \mu = 0, \\ \varphi\big((\tfrac{1}{\mu}\mathbb{1} - x)^{-1}\big), & 0 < |\mu| < \tfrac{1}{r} \end{cases}$$

defines a holomorphic function on $D^{-1} = \{\mu \in \mathbb{C} \,|\, |\mu| < \frac{1}{r}\}$ with Taylor expansion around zero given by

$$\sum_{n=0}^{\infty} \mu^{n+1}\varphi(x^n).$$

This expansion must converge on the whole disk D^{-1} and, in particular, for any $\mu \in D^{-1}$ we have

$$\lim_{n\to\infty} \mu^{n+1}\varphi(x^n) = 0.$$

Take now λ_0 such that $r < |\lambda_0| < \alpha(x)$. Then $\lambda_0^{-1} \in D^{-1}$ and

$$\lim_{n\to\infty} \lambda_0^{-n-1}\varphi(x^n) = 0.$$

Consider the family (sequence) of continuous functionals on $\mathsf{B}(\mathcal{H})^*$ given by

$$\mathsf{B}(\mathcal{H})^* \ni \varphi \longmapsto \lambda_0^{-n-1}\varphi(x^n) \in \mathbb{C}, \qquad n \in \mathbb{N}.$$

By the Banach-Steinhaus theorem there exists a constant $M < +\infty$ such that

$$\sup_{n\in\mathbb{N}} |\lambda_0|^{-n-1}\|x^n\| \le M.$$

Thus

$$\|x^n\| \le M|\lambda_0|^{n+1}, \qquad n \in \mathbb{N}$$

and consequently

$$\alpha(x) = \lim_{n\to\infty} \|x^n\|^{\frac{1}{n}} \le \lim_{n\to\infty} M^{\frac{1}{n}}|\lambda_0|^{1+\frac{1}{n}} = |\lambda_0| < \alpha(x).$$

---

[3]For $|\mu| > 1$ we have

$$\left|\sum_{n=0}^{\infty}(\lambda\mu)^{-n-1}\varphi(x^n)\right| \le \frac{\|\varphi\|}{|\mu|}\sum_{n=0}^{\infty}\frac{\|x^n\|}{|\lambda|^{n+1}}|\mu|^{-n} \le \frac{\|\varphi\|}{|\mu|}\sum_{n=0}^{\infty}\frac{\|x^n\|}{|\lambda|^{n+1}} \xrightarrow[|\mu|\to\infty]{} 0.$$

The series is convergent because $|\lambda| > \lim_{n\to\infty} \|x^n\|^{\frac{1}{n}}$, i.e. $|\lambda|^{-1}$ is strictly smaller than the radius of convergence $R = \big(\limsup_{n\to\infty} \|x^n\|^{\frac{1}{n}}\big)^{-1}$.

This contradiction shows that it is not possible to have $\alpha(x) > \sup\{|\lambda| \mid \lambda \in \sigma(x)\}$. In other words

$$\alpha(x) \leq \sup\{|\lambda| \mid \lambda \in \sigma(x)\},$$

which together with (1.4) gives $\alpha(x) = \sup\{|\lambda| \mid \lambda \in \sigma(x)\}$. □

For any $x \in \mathsf{B}(\mathcal{H})$ the quantity

$$\sup\{|\lambda| \mid \lambda \in \sigma(x)\}$$

is called the *spectral radius* of x and is denoted by $|\sigma(x)|$. Here are a few properties of the spectral radius:

- for any x we have $|\sigma(x)| \leq \|x\|$,
- if $x, y \in \mathsf{B}(\mathcal{H})$ commute then $|\sigma(xy)| \leq |\sigma(x)||\sigma(y)|$,
- $|\sigma(x^*)| = |\sigma(x)|$.

The last property follows immediately from the fact that

$$\sigma(x^*) = \overline{\sigma(x)} = \{\bar{\lambda} \mid \lambda \in \sigma(x)\}.$$

Proposition 1.8 *Let $x \in \mathsf{B}(\mathcal{H})$ and let $p(\lambda) = \alpha_0 + \alpha_1\lambda + \cdots + \alpha_n\lambda^n$ be a polynomial. Define*

$$p(x) = \alpha_0\mathbb{1} + \alpha_1 x + \cdots + \alpha_n x^n.$$

Then

$$\sigma\big(p(x)\big) = p\big(\sigma(x)\big) = \{p(\lambda) \mid \lambda \in \sigma(x)\}.$$

Proof

The statement is obvious for $n = 0$. Assume therefore that $n \geq 1$. Take $\lambda_0 \in \sigma(x)$, so that $\lambda_0\mathbb{1} - x$ is not invertible. Then the operator $p(\lambda_0)\mathbb{1} - p(x)$ cannot be invertible, as

$$\begin{aligned} p(\lambda_0)\mathbb{1} - p(x) &= \sum_{k=0}^{n} \alpha_k(\lambda_0^k\mathbb{1} - x^k) \\ &= \sum_{k=1}^{n} \alpha_k(\lambda_0^k\mathbb{1} - x^k) = (\lambda_0\mathbb{1} - x)\sum_{k=1}^{n}\alpha_k\sum_{j=1}^{k-1}\lambda_0^{k-j}x^{j-1}. \end{aligned}$$

and $\lambda_0\mathbb{1} - x$ and $\sum\limits_{k=1}^{n}\alpha_k\sum\limits_{j=1}^{k-1}\lambda_0^{k-j}x^{j-1}$ commute.

This shows that

$$p\big(\sigma(x)\big) \subset \sigma\big(p(x)\big).$$

On the other hand, if $\mu \notin \{p(\lambda) \mid \lambda \in \sigma(x)\}$ and $\lambda_1, \ldots, \lambda_n$ are the zeros of the polynomial $\mu - p(\lambda)$ then clearly $\lambda_1, \ldots, \lambda_n \notin \sigma(x)$. Furthermore

$$\mu - p(\lambda) = \gamma(\lambda_1 - \lambda)^{m_1} \cdots (\lambda_n - \lambda)^{m_n}$$

for some $m_1, \ldots, m_n \in \mathbb{N}$ and $\gamma \neq 0$, and therefore

$$\mu\mathbb{1} - p(x) = \gamma(\lambda_1\mathbb{1} - x)^{m_1} \cdots (\lambda_n\mathbb{1} - x)^{m_n}.$$

Thus $\mu\mathbb{1} - p(x)$ is invertible (as a product of invertible operators) and this means that $\mu \notin \sigma\big(p(x)\big)$. We have therefore shown that

$$p\big(\sigma(x)\big) \supset \sigma\big(p(x)\big).$$

□

1.3 Spectrum in C*-Algebras

Let A be a C*-algebra with unit (examples being $\mathsf{B}(\mathcal{H})$ for a Hilbert space $\mathcal{H}$ or $\mathrm{C}(X)$ for a compact space X, but these are far from exhaustive) and let $a \in \mathsf{A}$. Just as for elements of $\mathsf{B}(\mathcal{H})$ we say that a is *invertible* if there exists $b \in \mathsf{A}$ such that $ab = ba = \mathbb{1}$. We go on to define the *spectrum* of a:

$$\sigma(a) = \big\{\lambda \in \mathbb{C} \,\big|\, \text{the element } \lambda\mathbb{1} - a \text{ is not invertible}\big\}.$$

Consider for example the C*-algebra $\mathsf{A} = \mathrm{C}(X)$ with X a compact space. Then it is easy to see that the spectrum of an element $f \in \mathsf{A}$ coincides with the range of the function f.

Theorem 1.9
Let a be an element of a C-algebra with unit. Then*
(1) *$\sigma(a)$ is a non-empty compact subset of $\mathbb{C}$ contained in the disk with center 0 and radius $\|a\|$,*
(2) *the sequence $\big(\|a^n\|^{\frac{1}{n}}\big)_{n\in\mathbb{N}}$ is convergent and*

$$\lim_{n\to\infty} \|a^n\|^{\frac{1}{n}} = \sup\big\{|\lambda| \,\big|\, \lambda \in \sigma(a)\big\},$$

(Continued)

Theorem 1.9 (continued)

(3) *for any* $b \in \mathsf{A}$ *we have* $\sigma(ab) \cup \{0\} = \sigma(ba) \cup \{0\}$,

(4) *for any polynomial* $p \in \mathbb{C}[\cdot]$ *we have* $\sigma\big(p(a)\big) = p\big(\sigma(a)\big)$.

The proofs of all statements in the above theorem can be carried out by substituting A for $\mathsf{B}(\mathcal{H})$ in the proofs of Theorem 1.7 and Propositions 1.4 and 1.8.

Notes

Basics of the theory of operators on Hilbert spaces recalled in this chapter can be found in every textbook on the subject, including almost all items from the bibliography listed at the end of this book. More information on spectra of operators and elements of C*-algebras can be found e.g. in [Arv_2, Chapters 1 and 2], [Ped, Chapter 4], [Mau, Chapter VIII] as well as in many other books. A natural generalization of operator theory is provided by the theory of Banach algebras and in particular of C*-algebras ([Arv_1, Zel]). Within this approach many results can be obtained in a very elegant way, the price for this being paid by way of making the presentation more abstract and using a number of complicated structures. In this book, however, we aim to minimize the use of abstract theory reserving the more sophisticated path for further investigations on the part of the reader.

A wealth of great examples and exercises on the subject of spectra can be found in [Hal, Sections 9, 10 and 11] and in problem sections of textbooks such as [Arv_1, Arv_2, ReSi_1, Rud_2].

Continuous Functional Calculus

P. Sołtan, *A Primer on Hilbert Space Operators*, Compact Textbooks in Mathematics,
https://doi.org/10.1007/978-3-319-92061-0_2

In this chapter we will introduce by far the most important tool of the theory of operators on Hilbert space, namely *functional calculus* for self-adjoint operators. We begin with slightly more general considerations focused on normal operators which we will revisit later in ▶ Chap. 7.

An operator $x \in \mathrm{B}(\mathcal{H})$ is called *normal* if $x^*x = xx^*$. Equivalently x is normal if and only if for any $\xi \in \mathcal{H}$ we have $\|x\xi\| = \|x^*\xi\|$. Indeed: if $x^*x = xx^*$ then

$$\|x\xi\|^2 = \langle x\xi \,|\, x\xi\rangle = \langle \xi \,|\, x^*x\xi\rangle = \langle \xi \,|\, xx^*\xi\rangle = \langle x^*\xi \,|\, x^*\xi\rangle = \|x^*\xi\|^2.$$

Conversely if $\|x\xi\|^2 = \|x^*\xi\|^2$ for all $\xi \in \mathcal{H}$ then the sesquilinear forms

$$(\xi, \eta) \longmapsto \langle \xi \,|\, x^*x\eta\rangle \quad \text{and} \quad (\xi, \eta) \longmapsto \langle \xi \,|\, xx^*\eta\rangle$$

coincide on pairs of vectors of the form (ξ, ξ), so by the polarization identity they are equal. Consequently $x^*x = xx^*$.

An operator $x \in \mathrm{B}(\mathcal{H})$ is *self-adjoint* when $x = x^*$. Clearly, a self-adjoint operator is normal. Also any *unitary* operator, i.e. $u \in \mathrm{B}(\mathcal{H})$ satisfying $u^*u = uu^* = \mathbb{1}$, is normal.

Proposition 2.1 *Let $x \in \mathrm{B}(\mathcal{H})$ be a normal operator and let $\lambda \in \mathbb{C}$ and $\psi \in \mathcal{H}$ be such that $x\psi = \lambda\psi$.*[1] *Then $x^*\psi = \overline{\lambda}\psi$.*

Proof

One easily checks that the operator $\lambda\mathbb{1} - x$ is normal, and so

$$\|(\overline{\lambda}\mathbb{1} - x^*)\psi\| = \|(\lambda\mathbb{1} - x)^*\psi\| = \|(\lambda\mathbb{1} - x)\psi\| = 0.$$

□

[1] Restriction of a compact operator to an invariant subspace clearly is compact.

Corollary 2.2 *Let $x \in \mathsf{B}(\mathcal{H})$ be a normal operator and let $\lambda, \mu \in \sigma(x)$ be different eigenvalues of x. Then the eigenspaces of x for λ and μ are orthogonal.*

Proof

Let ψ and φ be eigenvectors of x for the eigenvalues λ and μ respectively: $x\psi = \lambda\psi$ and $x\varphi = \mu\varphi$. Using Proposition 2.1 we compute

$$\mu \langle \psi | \varphi \rangle = \langle \psi | x\varphi \rangle = \langle x^* \psi | \varphi \rangle = \langle \bar{\lambda}\psi | \varphi \rangle = \lambda \langle \psi | \varphi \rangle ,$$

which shows that $(\lambda - \mu) \langle \psi | \varphi \rangle = 0$. As $\lambda \neq \mu$, we must have $\langle \psi | \varphi \rangle = 0$. □

In particular eigenspaces of a self-adjoint operator for different eigenvalues also have to be orthogonal.

Proposition 2.3 *Let $x \in \mathsf{B}(\mathcal{H})$. Then*
(1) *if x is normal then $|\sigma(x)| = \|x\|$,*
(2) *if x is self-adjoint then $\sigma(x) \subset \mathbb{R}$.*

Proof

Assume first that x is normal. We have

$$\begin{aligned} \|x^2\|^2 = \big\|(x^2)^*(x^2)\big\| &= \|x^*x^*xx\| \\ &= \|x^*xx^*x\| = \big\|(x^*x)^*(x^*x)\big\| = \|x^*x\|^2 = \|x\|^4, \end{aligned}$$

so that $\|x^2\| = \|x\|^2$. Since for any $k \in \mathbb{N}$ the operator x^{2^k} is normal as well, for any $n \in \mathbb{N}$ we have

$$\|x^{2^n}\| = \big\|(x^{2^{n-1}})^2\big\| = \|x^{2^{n-1}}\|^2 = \|x^{2^{n-2}}\|^4 = \cdots = \|x\|^{2^n}.$$

In particular, the sequence $\big(\|x^n\|^{\frac{1}{n}}\big)_{n \in \mathbb{N}}$ has a subsequence which is constant and equal to $\|x\|$. On the other hand

$$|\sigma(x)| = \lim_{n\to\infty} \|x^n\|^{\frac{1}{n}}.$$

Now let x be self-adjoint. Take any $\lambda \in \sigma(x)$ and write $\lambda = \alpha + \mathrm{i}\beta$ with $\alpha, \beta \in \mathbb{R}$. For $n \in \mathbb{N}$ consider the operator $x_n = x - (\alpha - \mathrm{i}n\beta)\mathbb{1}$. We have $\mathrm{i}(n+1)\beta \in \sigma(x_n)$, so that

$$\begin{aligned} (n+1)^2\beta^2 \le |\sigma(x_n)|^2 \le \|x_n\|^2 &= \|{x_n}^* x_n\| \\ &= \big\|\big(x - (\alpha + \mathrm{i}n\beta)\mathbb{1}\big)\big(x - (\alpha - \mathrm{i}n\beta)\mathbb{1}\big)\big\| \\ &= \big\|(x - \alpha\mathbb{1})^2 + n^2\beta^2\mathbb{1}\big\| \\ &\le \big\|(x - \alpha\mathbb{1})^2\big\| + n^2\beta^2 \end{aligned}$$

for all n. This means that $\beta = 0$ and consequently $\lambda \in \mathbb{R}$. □

Before stating the main theorem of this chapter let us recall that a $*$-*isomorphism* is a bijective map between algebras with involution preserving addition, multiplication and involution.

Theorem 2.4 (Continuous Functional Calculus)
Let $x \in \mathsf{B}(\mathcal{H})$ be self-adjoint. Then there exists a unique map $\mathsf{C}(\sigma(x)) \to \mathsf{B}(\mathcal{H})$ denoted by

$$\mathsf{C}(\sigma(x)) \ni f \longmapsto f(x) \in \mathsf{B}(\mathcal{H})$$

such that
- *if f is a polynomial function $f(\lambda) = \alpha_0 + \alpha_1\lambda + \cdots + \alpha_n\lambda^n$ then*

$$f(x) = \alpha_0 \mathbb{1} + \alpha_1 x + \cdots + \alpha_n x^n,$$

- *$\|f(x)\| = \|f\|_\infty$ for all $f \in \mathsf{C}(\sigma(x))$.*

Moreover the map $f \mapsto f(x)$ is a $*$-isomorphism of the C^**-algebra $\mathsf{C}(\sigma(x))$ onto* $\mathrm{C}^*(x, \mathbb{1})$.

Proof
Let $\mathcal{P}(\sigma(x))$ be the set of polynomial functions on $\sigma(x)$, i.e. restrictions of polynomials to $\sigma(x)$ and let $\Psi : \mathbb{C}[\,\cdot\,] \to \mathsf{C}(\sigma(x))$ be the map $p \mapsto p|_{\sigma(x)}$. For a polynomial $p \in \mathbb{C}[\,\cdot\,]$ we have

$$\begin{aligned}\|p(x)\| = |\sigma(p(x))| &= \sup\{|\mu| \,|\, \mu \in \sigma(p(x))\} \\ &= \sup\{|p(\lambda)| \,|\, \lambda \in \sigma(x)\} = \|\Psi(p)\|_\infty.\end{aligned}$$

Thus the mapping

$$\mathbb{C}[\,\cdot\,] \ni p \longmapsto p(x) \in \mathsf{B}(\mathcal{H})$$

factorizes through Ψ, i.e. there exists a linear $\Phi : \mathcal{P}(\sigma(x)) \to \mathsf{B}(\mathcal{H})$ such that

$$p(x) = \Phi(\Psi(p)).$$

Moreover Φ is an isometry.

The space $\mathcal{P}(\sigma(x))$ is dense in $\mathsf{C}(\sigma(x))$ and its image under Φ is dense in the C^*-algebra $\mathrm{C}^*(x, \mathbb{1})$. Therefore Φ extends uniquely to an isometry from $\mathsf{C}(\sigma(x))$ onto $\mathrm{C}^*(x, \mathbb{1})$ which we will from now on denote by $f \mapsto f(x)$. It is clear that this map satisfies

- if f is a polynomial function $f(\lambda) = \alpha_0 + \alpha_1\lambda + \cdots + \alpha_n\lambda^n$ then

$$f(x) = \alpha_0\mathbb{1} + \alpha_1 x + \cdots + \alpha_n x^n,$$

- $\|f(x)\| = \|f\|_\infty$ for all $f \in \mathrm{C}(\sigma(x))$,

and that these conditions determine this map uniquely.

The properties

$$\begin{aligned}
(f+g)(x) &= f(x) + g(x), && f, g \in \mathrm{C}(\sigma(x)),\\
(fg)(x) &= f(x)g(x), && f, g \in \mathrm{C}(\sigma(x)),\\
(\lambda f)(x) &= \lambda f(x), && \lambda \in \mathbb{C},\ f \in \mathrm{C}(\sigma(x)),\\
\overline{f}(x) &= f(x)^*, && f \in \mathrm{C}(\sigma(x))
\end{aligned}$$

are easily checked on elements of $\mathcal{P}(\sigma(x))$ and for general continuous functions on $\sigma(x)$ we use approximation by polynomial functions. In other words $f \mapsto f(x)$ is an isometric $*$-isomorphism of $\mathrm{C}(\sigma(x))$ onto $\mathrm{C}^*(x, \mathbb{1})$. □

Remark 2.5

(1) Let $x \in \mathrm{B}(\mathcal{H})$ be self-adjoint and let f be a continuous function on $\sigma(x)$ which can be uniformly approximated by polynomials without constant term. Then $f(x)$ belongs to $\mathrm{C}^*(x)$ which can be strictly smaller than $\mathrm{C}^*(x, \mathbb{1})$.

(2) If $0 \notin \sigma(x)$ (i.e. x is invertible) then any $f \in \mathrm{C}(\sigma(x))$ can be uniformly approximated by polynomials without constant term. In particular if x is an invertible self-adjoint operator then the constant function 1 on $\sigma(x)$ can be uniformly approximated by polynomials without constant term. Therefore $\mathbb{1}$ belongs to $\mathrm{C}^*(x)$ and consequently $\mathrm{C}^*(x, \mathbb{1}) = \mathrm{C}^*(x)$. Moreover, also the function

$$\iota : \sigma(x) \ni \lambda \longmapsto \lambda^{-1} \in \mathbb{C}$$

can be uniformly approximated by polynomials without constant term. It is easy to see that $\iota(x) = x^{-1}$, so $x^{-1} \in \mathrm{C}^*(x)$.

(3) Even when x is not self-adjoint, but $0 \notin \sigma(x)$ we can show that $x^{-1} \in \mathrm{C}^*(x)$. Indeed: the operator x^*x is self-adjoint and invertible with inverse $x^{-1}(x^{-1})^*$. It follows from (2) that $x^{-1}(x^{-1})^*$ belongs to $\mathrm{C}^*(x^*x) \subset \mathrm{C}^*(x)$. Therefore

$$x^{-1} = \left(x^{-1}(x^{-1})^*\right)x^* \in \mathrm{C}^*(x).$$

Proposition 2.6 *Let x and y be commuting self-adjoint operators on $\mathcal{H}$. Then for any $f \in \mathrm{C}(\sigma(x))$ and $g \in \mathrm{C}(\sigma(y))$ we have $f(x)g(y) = g(y)f(x)$.*

Proof
Let $(f_n)_{n\in\mathbb{N}}$ and $(g_n)_{n\in\mathbb{N}}$ be sequences of polynomials uniformly approximating f and g on $\sigma(x)$ and $\sigma(y)$ respectively. It is clear that for any n we have $f_n(x)g_n(y) = g_n(y)f_n(x)$ and

$$f_n(x)g_n(y) \xrightarrow[n\to\infty]{} f(x)g(y) \quad \text{and} \quad g_n(y)f_n(x) \xrightarrow[n\to\infty]{} g(y)f(x).$$

□

Remark 2.7 Let $x, y \in \mathrm{B}(\mathcal{H})$ and assume the $x = x^*$ and $xy = yx$. Then for any $f \in \mathrm{C}(\sigma(x))$ the operator $f(x)$ also commutes with y. The proof follows the lines of the proof of Proposition 2.6 .

Theorem 2.8 (Spectral Mapping Theorem)
Let $x \in \mathrm{B}(\mathcal{H})$ be self-adjoint. Then for any $f \in \mathrm{C}(\sigma(x))$ we have

$$\sigma(f(x)) = f(\sigma(x)).$$

Proof
For any μ

$$\mu\mathbb{1} - f(x) = (\mu - f)(x).$$

Therefore $\mu\mathbb{1} - f(x)$ is invertible if and only if the function $\mu - f$ is an invertible element of $\mathrm{C}(\sigma(x))$, i.e. is and only if $\mu \notin f(\sigma(x))$. □

Remark 2.9 Let x be a self-adjoint operator and assume that $x^n = 0$ for some $n \in \mathbb{N}$. Then $\sigma(x) = \{0\}$, because $\sigma(x^n) = \{0\}$ and $\sigma(x^n) = \{\lambda^n \mid \lambda \in \sigma(x)\}$. It follows that $\|x\| = |\sigma(x)| = 0$, i.e. $x = 0$.

Proposition 2.10 *Let $x \in \mathrm{B}(\mathcal{H})$ be self-adjoint. Then for any real-valued $g \in \mathrm{C}(\sigma(x))$ the operator $g(x)$ is self-adjoint and for any $f \in \mathrm{C}(g(\sigma(x)))$ we have*

$$f(g(x)) = (f \circ g)(x).$$

Proof
Since $\mathrm{C}(\sigma(x)) \ni g \mapsto g(x) \in \mathrm{C}^*(x, \mathbb{1})$ is a $*$-isomorphism, $g = \overline{g}$ implies $g(x)^* = g(x)$.
So let us fix a real-valued $g \in \mathrm{C}(\sigma(x))$ and consider the mapping

$$\mathrm{C}(g(\sigma(x))) \ni f \longmapsto (f \circ g)(x) \in \mathrm{B}(\mathcal{H}).$$

It is easy to see that it is isometric and maps polynomial functions onto corresponding polynomials in $g(x)$. By uniqueness of the continuous functional calculus it must be equal to the map $f \mapsto f\big(g(x)\big)$. □

Notes

It is difficult to find a more fundamental concept in the theory of operators on Hilbert spaces than functional calculus and, in particular, the continuous functional calculus. As we already mentioned earlier, it is the first step in the development of the theory of C*-algebras and general operator algebras. The reader will find plentiful exercises and examples e.g. in [Arv_1, Chapter 1], [Arv_2, Chapter 2], [Hal, Section 15], [Ped, Sections 4.4 and 4.5], [ReSi_1, Chapter VII], [Rud_2, Chapters 10 and 12].

Let us also note that continuous functional calculus for self-adjoint operators can be extended to a larger class of operators (namely the so called *normal* operators, see ▶ Sect. 4.6). We will do this in ▶ Sect. 7.4.

Positive Operators

P. Sołtan, *A Primer on Hilbert Space Operators*, Compact Textbooks in Mathematics,
https://doi.org/10.1007/978-3-319-92061-0_3

One of the first and, incidentally, very effective applications of functional calculus is to take square roots of positive operators. In this chapter we will introduce positive operators and related notions of a projection and a partial isometry. We will prove existence and uniqueness of polar decomposition of bounded operators. We will also briefly investigate the partial order on operators defined by the notion of a positive operator. In particular we will show that any bounded monotonically increasing net of self-adjoint operators has a supremum which is also the limit of the net in *strong topology*.

3.1 Positive Operators

An operator $x \in \mathsf{B}(\mathcal{H})$ is called *positive* if x is self-adjoint and $\sigma(x) \subset \mathbb{R}_+$. In this case we write $x \geq 0$. The set of positive operators on $\mathcal{H}$ will be denoted by the symbol $\mathsf{B}(\mathcal{H})_+$.

Theorem 3.1
Let $x \in \mathsf{B}(\mathcal{H})$ be positive. Then there exists a unique positive $a \in \mathsf{B}(\mathcal{H})$ such that $a^2 = x$.

Proof
The function $f : \lambda \mapsto \lambda^{\frac{1}{2}}$ is continuous on $\sigma(x)$. Put $a = f(x)$. Clearly

$$a^2 = f(x)f(x) = x,$$

since $f(\lambda)f(\lambda) = \lambda$ for all $\lambda \in \sigma(x)$. Moreover $\sigma(a) = f\big(\sigma(x)\big)$ is contained in $\mathbb{R}_+$, so that $a \in \mathsf{B}(\mathcal{H})_+$.

Now let b be a positive operator such that $b^2 = x$. By Proposition 2.10 we have $b = f(g(b))$, where $g(\lambda) = \lambda^2$ for all $\lambda \in \mathbb{R}_+$. Therefore

$$b = f(b^2) = f(x) = a.$$

□

Let $x \in \mathsf{B}(\mathcal{H})$ be a positive operator. The unique positive operator a such that $a^2 = x$ is called the *square root* of x. It follows from the existence of square roots that any positive operator is the square of a self-adjoint operator. Moreover, by the spectral mapping theorem, the square of a self-adjoint operator y is positive, since its spectrum consists of squares of elements of $\sigma(y)$.

Proposition 3.2 *Let $x \in \mathsf{B}(\mathcal{H})$ be self-adjoint. Then there exists a unique pair (a, b) of positive operators such that*

$$x = a - b \quad \text{and} \quad ab = 0. \tag{3.1}$$

Proof

Consider the following two continuous functions on $\sigma(x)$:

$$f(\lambda) = \begin{cases} \lambda, & \lambda \geq 0, \\ 0, & \lambda < 0 \end{cases} \quad \text{and} \quad g(\lambda) = \begin{cases} 0, & \lambda > 0, \\ -\lambda, & \lambda \leq 0. \end{cases}$$

It is check that $a = f(x)$ and $b = g(x)$ satisfy the conditions (3.1).

On the other hand if a pair (a, b) of positive operators satisfies (3.1) then first of all a and b must commute:

$$ba = b^*a^* = (ab)^* = 0^* = 0 = ab.$$

It follows that also $a^{\frac{1}{2}}$ and $b^{\frac{1}{2}}$ commute (Proposition 2.6) and, moreover, $a^{\frac{1}{2}}b^{\frac{1}{2}} = 0$, since

$$\left(a^{\frac{1}{2}}b^{\frac{1}{2}}\right)^2 = ab = 0$$

(cf. Remark 2.9).

From this we obtain

$$\left(a^{\frac{1}{2}} + b^{\frac{1}{2}}\right)^2 = a + b,$$

so $a + b$ is positive, as a square of a self-adjoint operator. Furthermore

$$(a + b)^2 = (a - b)^2 = x^2,$$

which means that $a+b$ is the unique square root of the positive operator x^2, i.e. $a+b = (x^2)^{\frac{1}{2}}$. Consequently

$$a = \tfrac{1}{2}\big((a+b)+(a-b)\big) = \tfrac{1}{2}\big((x^2)^{\frac{1}{2}} + x\big) = f(x)$$

and similarly

$$b = \tfrac{1}{2}\big((a+b)-(a-b)\big) = \tfrac{1}{2}\big((x^2)^{\frac{1}{2}} - x\big) = g(x).$$

□

Operators a and b defined in Proposition 3.2 are called the *positive part* and *negative part* of the self-adjoint operator x. We denote them by the symbols x^+ and x^- respectively. Note that x^+ and x^- belong to $C^*(x)$.

Proposition 3.3 *Let* $x \in B(\mathcal{H})$. *Then*
(1) $x = 0$ *if and only if for any* $\xi \in \mathcal{H}$ *we have* $\langle \xi \,|\, x\xi \rangle = 0$,
(2) $x = x^*$ *if and only if for any* $\xi \in \mathcal{H}$ *we have* $\langle \xi \,|\, x\xi \rangle \in \mathbb{R}$.

Proof
Ad (1). It follows from the polarization formula that if $\langle \xi \,|\, x\xi \rangle = 0$ for all ξ then the sesquilinear form

$$(\xi, \eta) \longmapsto \langle \xi \,|\, x\eta \rangle\,, \qquad \xi, \eta \in \mathcal{H}$$

is the zero form, and so $x = 0$.

Ad (2). Consider two sesquilinear forms:

$$F_1 : (\xi, \eta) \longmapsto \langle \xi \,|\, x\eta \rangle \quad \text{and} \quad F_2 : (\xi, \eta) \longmapsto \langle x\xi \,|\, \eta \rangle\,.$$

We have

$$F_1(\xi, \xi) = \langle \xi \,|\, x\xi \rangle = \overline{\langle \xi \,|\, x\xi \rangle} = \langle x\xi \,|\, \xi \rangle = F_2(\xi, \xi),$$

because $\langle \xi \,|\, x\xi \rangle \in \mathbb{R}$. It follows that $F_1 = F_2$, so that x is self-adjoint. □

Proposition 3.4 *Let* $x \in B(\mathcal{H})$. *Then the following conditions are equivalent:*
(1) x *is positive,*
(2) *there exists a self-adjoint operator* y *such that* $x = y^2$,
(3) *there exists an operator* z *such that* $x = z^* z$,
(4) *for any* $\xi \in \mathcal{H}$ *we have* $\langle \xi \,|\, x\xi \rangle \geq 0$.

Proof

(1) $\Rightarrow$ (2) follows from Theorem 3.1: $y = x^{\frac{1}{2}}$. The implication (2) $\Rightarrow$ (3) is obvious and so is (3) $\Rightarrow$ (4): for any ξ we have

$$\langle \xi \,|\, x\xi \rangle = \langle \xi \,|\, z^*z\xi \rangle = \langle z\xi \,|\, z\xi \rangle = \|z\xi\|^2 \geq 0.$$

(4) $\Rightarrow$ (1). Decompose x into its positive and negative part, $x = x^+ - x^-$, and let $f(\lambda) = \lambda^{\frac{3}{2}}$ for $\lambda \in \mathbb{R}_+$. Since $x^+x^- = 0$, for any ξ we have

$$\begin{aligned} 0 \leq \langle x^-\xi \,|\, x(x^-\xi) \rangle &= \langle \xi \,|\, (x^-xx^-)\xi) \rangle = \langle \xi \,|\, (x^-(x^+ - x^-)x^-)\xi) \rangle \\ &= -\left\langle \xi \,\middle|\, (x^-)^3\xi \right\rangle = -\langle x^-\xi \,|\, x^-(x^-\xi) \rangle \\ &= -\left\langle x^-\xi \,\middle|\, (x^-)^{\frac{1}{2}}(x^-)^{\frac{1}{2}}(x^-\xi) \right\rangle \\ &= -\langle f(x^-)\xi \,|\, f(x^-)\xi \rangle \leq 0. \end{aligned}$$

It follows that $\langle \xi \,|\, (x^-)^3\xi \rangle = 0$ for all ξ, which by Proposition 3.3(1) means that $(x^-)^3 = 0$. It follows now from Remark 2.9 that $x^- = 0$. In particular $x = x^+ \geq 0$. □

In the terminology of quantum physics, for $x \in \mathsf{B}(\mathcal{H})$ the quantity $\langle \xi \,|\, x\xi \rangle$ is referred to as the *expectation value* of x in the state determined by ξ (albeit this terminology usually applies only to vectors ξ of norm 1). Proposition 3.4 says, among other things, that an operator is positive if and only if its expectation values in all states are positive.

The notion of positivity for elements of $\mathsf{B}(\mathcal{H})$ allows us to define a partial order relation of $\mathsf{B}(\mathcal{H})$: let $x, y \in \mathsf{B}(\mathcal{H})$. We say that x *dominates* y if $x - y \geq 0$. We write this as $x \geq y$. Let us note for future reference that for any $x \in \mathsf{B}(\mathcal{H})$ we have $x^*x \leq \|x\|^2\mathbb{1}$, The proof is based on a simple computation:

$$\langle \xi \,|\, x^*x\xi \rangle = \|x\xi\|^2 \leq \|x\|^2\|\xi\|^2 = \langle \xi \,|\, \|x\|^2\xi \rangle, \qquad \xi \in \mathcal{H}.$$

Moreover, if $x^*x \leq \mathbb{1}$, then by the same computation $\|x\| \leq 1$. In other words we have the following:

Proposition 3.5 *For any positive* $t \in \mathsf{B}(\mathcal{H})$

$$\big(t \leq \mathbb{1}\big) \Longleftrightarrow \big(\|t\| \leq 1\big).$$

Proposition 3.4 can be used to obtain information about the partial order on $\mathsf{B}(\mathcal{H})$ both of geometric and algebraic nature:

Corollary 3.6 $\mathsf{B}(\mathcal{H})_+$ *is a convex cone, i.e.*

(1) *if* $x \in \mathsf{B}(\mathcal{H})_+$ *and* $\lambda \in \mathbb{R}_+$ *then* $\lambda x \in \mathsf{B}(\mathcal{H})_+$
(2) *if* $x, y \in \mathsf{B}(\mathcal{H})_+$ *then* $x + y \in \mathsf{B}(\mathcal{H})_+$,
(3) $\mathsf{B}(\mathcal{H})_+ \cap \big(-\mathsf{B}(\mathcal{H})_+\big) = \{0\}$.

Moreover, if $x \in \mathsf{B}(\mathcal{H})_+$ *and* $y \in \mathsf{B}(\mathcal{H})$ *then* $y^*xy \in \mathsf{B}(\mathcal{H})_+$.

Proof
The first of the listed properties of $\mathsf{B}(\mathcal{H})_+$ follows immediately from the definitions, while the second is a consequence of characterization of positivity given by point (4) of Proposition 3.4. If $x \in \mathsf{B}(\mathcal{H})_+ \cap \big(-\mathsf{B}(\mathcal{H})_+\big)$ then $\sigma(x) = \{0\}$, and since x is self-adjoint, we get $x = 0$.

Let $x \in \mathsf{B}(\mathcal{H})_+$ and $y \in \mathsf{B}(\mathcal{H})$. Then for any $\xi \in \mathcal{H}$ we have

$$\left\langle \xi \,\middle|\, y^* x y \xi \right\rangle = \langle y\xi \,|\, x(y\xi)\rangle \geq 0,$$

which means that $y^* x y \geq 0$. □

Let us note here that functional calculus for self-adjoint operators is compatible with the order structure on $\mathsf{B}(\mathcal{H})$ in the sense that if $x \in \mathsf{B}(\mathcal{H})$ is self-adjoint and $f, g \in \mathsf{C}\big(\sigma(x)\big)$ satisfy $f \geq g$ then $f(x) \geq g(x)$ in $\mathsf{B}(\mathcal{H})$ because

$$f(x) - g(x) = (f - g)(x) = \big((f - g)^{\frac{1}{2}}(x)\big)^* \big((f - g)^{\frac{1}{2}}(x)\big) \in \mathsf{B}(\mathcal{H})_+.$$

Proposition 3.7 *Let $x, y \in \mathsf{B}(\mathcal{H})_+$ be such that $0 \leq x \leq y$ and assume that x is invertible. Then y is invertible and $y^{-1} \leq x^{-1}$.*

Proof
The compact set $\sigma(x)$ lies in $[0, +\infty[$ and it does not contain 0. Therefore there exists $\delta > 0$ such that $\sigma(x) \subset [\delta, +\infty[$. By functional calculus we immediately get $x \geq \delta\mathbb{1}$, and so $y \geq \delta\mathbb{1}$. Thus, again by functional calculus, we find that y is invertible.

Now multiplying both sides of the inequality $y \geq x$ by $y^{-\frac{1}{2}}$ from left and right and using Corollary 3.6 we obtain

$$y^{-\frac{1}{2}} x y^{-\frac{1}{2}} \leq \mathbb{1}$$

which, in particular, means that $\big\| y^{-\frac{1}{2}} x y^{-\frac{1}{2}} \big\| \leq 1$. Thus

$$\big\| x^{\frac{1}{2}} y^{-\frac{1}{2}} \big\|^2 = \big\| y^{-\frac{1}{2}} x y^{-\frac{1}{2}} \big\| \leq 1$$

and consequently

$$\big\| y^{-\frac{1}{2}} x^{\frac{1}{2}} \big\| = \big\| \big(x^{\frac{1}{2}} y^{-\frac{1}{2}}\big)^* \big\| \leq 1.$$

This, in turn, shows that

$$\big(y^{-\frac{1}{2}} x^{\frac{1}{2}}\big)^* \big(y^{-\frac{1}{2}} x^{\frac{1}{2}}\big) \leq \big\| y^{-\frac{1}{2}} x^{\frac{1}{2}} \big\|^2 \mathbb{1} \leq \mathbb{1}$$

or in other words

$$x^{\frac{1}{2}} y^{-1} x^{\frac{1}{2}} \leq \mathbb{1}.$$

Multiplying both sides of this inequality by $x^{-\frac{1}{2}}$ from left and right we arrive at $y^{-1} \leq x^{-1}$.

□

Taking roots of positive operators has many other applications. An example of such an application is provided by the following useful proposition:

Proposition 3.8 *Any $x \in \mathsf{B}(\mathcal{H})$ is a linear combination of four unitary operators.*

Proof
Writing

$$x = \tfrac{1}{2}(x + x^*) + \mathrm{i}\tfrac{1}{2\mathrm{i}}(x - x^*),$$

we express x as a linear combination of self-adjoint operators. Now any self-adjoint operator $y \in \mathsf{B}(\mathcal{H})$ of norm 1 can be written in the form

$$y = \tfrac{1}{2}\Big(\big(y + \mathrm{i}(\mathbb{1} - y^2)^{\frac{1}{2}}\big) + \big(y - \mathrm{i}(\mathbb{1} - y^2)^{\frac{1}{2}}\big)\Big).$$

The operators $y \pm \mathrm{i}(\mathbb{1} - y^2)^{\frac{1}{2}}$ are unitary, since

$$\begin{aligned}\big(y \pm \mathrm{i}(\mathbb{1} - y^2)^{\frac{1}{2}}\big)^*&\big(y \pm \mathrm{i}(\mathbb{1} - y^2)^{\frac{1}{2}}\big)\\ &= \big(y \mp \mathrm{i}(\mathbb{1} - y^2)^{\frac{1}{2}}\big)\big(y \pm \mathrm{i}(\mathbb{1} - y^2)^{\frac{1}{2}}\big) = y^2 + (\mathbb{1} - y^2) = \mathbb{1}\end{aligned}$$

and similarly $\big(y \pm \mathrm{i}(\mathbb{1} - y^2)^{\frac{1}{2}}\big)\big(y \pm \mathrm{i}(\mathbb{1} - y^2)^{\frac{1}{2}}\big)^* = \mathbb{1}$.

□

3.2 Projections

A *projection* is an operator $e \in \mathsf{B}(\mathcal{H})$ such that $e = e^2$ and $e = e^*$. We are using the term "projection" in a way which is slightly more restrictive than usual. More precisely, the term "projection" is often taken to mean "idempotent operator", i.e. an $x \in \mathsf{B}(\mathcal{H})$ such that $x^2 = x$. We have chosen to use the more restrictive definition requiring that all projection be self-adjoint, because the extra flexibility of working with general idempotents will not have any significance for the topics considered in this book. We leave it to the reader to check that an idempotent operator is self-adjoint if and only if its kernel and range are orthogonal.

Any projection e is a positive operator of norm 1 or 0. Moreover the operator $\mathbb{1} - e$ is also a projection. The set of all projections in $\mathsf{B}(\mathcal{H})$ will be denoted by $\mathrm{Proj}\big(\mathsf{B}(\mathcal{H})\big)$.

Note that an operator $e \in \mathsf{B}(\mathcal{H})$ is a projection if and only if it is self-adjoint and the two functions $\lambda \mapsto \lambda^2$ and $\lambda \mapsto \lambda$ coincide on $\sigma(e)$. It follows that $\sigma(e) \subset \{0, 1\}$. Conversely, if $e \in \mathsf{B}(\mathcal{H})$ is self adjoint and $\sigma(e) \subset \{0, 1\}$ then $e^2 = e$ precisely because the functions $\lambda \mapsto \lambda^2$ and $\lambda \mapsto \lambda$ are equal on $\sigma(e)$.

Let $e \in \mathsf{B}(\mathcal{H})$ be a projection. Then the subspace $e\mathcal{H} \subset \mathcal{H}$ is closed and $(\mathbb{1} - e)\mathcal{H}$ is the orthogonal complement of $e\mathcal{H}$. Of course for any closed subspace $\mathcal{S} \subset \mathcal{H}$ there exists a unique projection $e \in \mathsf{B}(\mathcal{H})$ such that $\mathcal{S} = e\mathcal{H}$. It is not hard to check that if e and f are projections then $e \leq f$ if and only if $e\mathcal{H} \subset f\mathcal{H}$.

We say that two projections $e_1, e_2 \in \mathsf{B}(\mathcal{H})$ are *orthogonal* if $e_1 e_2 = 0$. This is equivalent to the subspaces $e_1\mathcal{H}$ and $e_2\mathcal{H}$ being orthogonal. Let us also note that for any two projections e and f the operator $e + f$ is a projection if and only if e and f are orthogonal.

With any operator $x \in \mathsf{B}(\mathcal{H})$ we can associate two projections $\mathbf{l}(x)$ and $\mathbf{r}(x)$: $\mathbf{l}(x)$ is defined to be the projection onto the closure of $x\mathcal{H}$, while $\mathbf{r}(x)$ is the projection onto the orthogonal complement of $\ker x$. We have

$$\mathbf{l}(x)x = x = x\mathbf{r}(x).$$

The projections $\mathbf{l}(x)$ and $\mathbf{r}(x)$ are respectively called the *left support* and the *right support* of x.

Since for any $x \in \mathsf{B}(\mathcal{H})$ we have $(x\mathcal{H})^{\perp} = \ker x^*$ ($\langle \eta | x\xi \rangle = 0$ for all ξ if and only if $\langle x^*\eta | \xi \rangle = 0$ for all ξ), it is easy to see that $\mathbf{l}(x) = \mathbf{r}(x^*)$ (equivalently $\mathbf{l}(x^*) = \mathbf{r}(x)$ for all x). In particular if $x = x^*$ then $\mathbf{l}(x) = \mathbf{r}(x)$ and, in this case, the common value of $\mathbf{l}(x)$ and $\mathbf{r}(x)$ is denoted by $\mathbf{s}(x)$ and called the *support* of x.

Lemma 3.9 *For any $x \in \mathsf{B}(\mathcal{H})$ we have* $\ker x = \ker x^*x$.

Proof

It is clear that $\ker x \subset \ker x^*x$. On the other hand, if $\xi \in \ker x^*x$ then $\langle \xi | x^*x\xi \rangle = 0$, so $\|x\xi\|^2 = 0$. □

Proposition 3.10 *For $x \in \mathsf{B}(\mathcal{H})$ we have $\mathbf{r}(x) = \mathbf{s}(x^*x)$ and $\mathbf{l}(x) = \mathbf{s}(xx^*)$.*

Proof

Using Lemma 3.9 we see that

$$\begin{aligned} \mathbf{r}(x) &= \text{projection onto } (\ker x)^{\perp} \\ &= \text{projection onto } (\ker x^*x)^{\perp} = \mathbf{r}(x^*x) = \mathbf{s}(x^*x) \end{aligned}$$

and similarly

$$\begin{aligned} \mathbf{l}(x) &= \text{projection onto } \overline{x\mathcal{H}} = \text{projection onto } (\ker x^*)^{\perp} \\ &= \text{projection onto } (\ker xx^*)^{\perp} = \mathbf{r}(xx^*) = \mathbf{s}(xx^*). \end{aligned}$$

□

3.3 Partial Isometries

Proposition 3.11 *Let $v \in \mathsf{B}(\mathcal{H})$. Then the following conditions are equivalent:*
(1) v^*v *is a projection,*
(2) $vv^*v = v$,
(3) $v^*vv^* = v^*$,
(4) vv^* *is a projection.*

Proof
(1) and (4) are equivalent by Proposition 1.4 and remarks at the beginning of ▶ Sect. 3.2. More precisely $\sigma(v^*v)$ is contained in $\{0, 1\}$ if and only if so is $\sigma(vv^*)$, and furthermore both v^*v and vv^* are self-adjoint. It follows that v^*v is a projection if and only if vv^* is a projection.

(1) ⇒ (2): the support of any projection e is equal to e. Therefore, setting $e = v^*v$ we obtain

$$v^*v = \mathbf{s}(v^*v) = \mathbf{r}(v)$$

by Proposition 3.10. In particular $vv^*v = v$.

(2) ⇒ (3): taking adjoints of both sides of the equality $vv^*v = v$ we arrive at $v^*vv^* = v^*$.

(3) ⇒ (4): multiplying both sides of $v^*vv^* = v^*$ by v from the left we get $vv^*vv^* = vv^*$, i.e. the self-adjoint operator vv^* satisfies $(vv^*)^2 = vv^*$. □

An operator $v \in \mathsf{B}(\mathcal{H})$ is called a *partial isometry* if it satisfies the equivalent conditions of Proposition 3.11. The projections v^*v and vv^* are referred to as the *initial projection* and *final projection* of the partial isometry v, while their ranges $v^*v\mathcal{H}$ and $vv^*\mathcal{H}$ are respectively the *initial subspace* and *final subspace* of v.

Proposition 3.12 *An operator $v \in \mathsf{B}(\mathcal{H})$ is a partial isometry if and only if there exists a closed subspace $\mathcal{S} \subset \mathcal{H}$ such that*

$$\|v\xi\| = \begin{cases} \|\xi\|, & \xi \in \mathcal{S}, \\ 0, & \xi \in \mathcal{S}^{\perp}. \end{cases} \tag{3.2}$$

In this case $\mathcal{S}$ is the initial subspace of v and the final subspace of v is $v\mathcal{S}$.

Proof
Assume first that v is a partial isometry and let $\mathcal{S} = v^*v\mathcal{H}$. Then $\mathcal{S}$ is a closed subspace and for $\xi \in \mathcal{S}$

$$\langle v\xi \,|\, v\xi\rangle = \langle \xi \,|\, v^*v\xi\rangle = \langle \xi \,|\, \xi\rangle ,$$

while for $\xi \in \mathcal{S}^\perp$

$$\langle v\xi \,|\, v\xi\rangle = \langle \xi \,|\, v^* v\xi\rangle = 0.$$

In other words we obtain (3.2).

Conversely, if $v \in \mathrm{B}(\mathcal{H})$ is an operator satisfying (3.2) for some closed subspace $\mathcal{S} \subset \mathcal{H}$ then clearly $v^*v\xi = 0$ for $\xi \in \mathcal{S}^\perp$. Moreover for $\xi \in \mathcal{S}$ we have $v^*v\xi = \xi$. Indeed: take any $\eta \in \mathcal{H}$ and let $\eta = \eta_{\mathcal{S}} + \eta_{\mathcal{S}^\perp}$ be the decomposition of η into components in $\mathcal{S}$ and in $\mathcal{S}^\perp$. Then

$$\begin{aligned}\langle \eta | v^* v\xi\rangle &= \langle \eta_{\mathcal{S}} + \eta_{\mathcal{S}^\perp} | v^* v\xi\rangle = \langle v\eta_{\mathcal{S}} + v\eta_{\mathcal{S}^\perp} | v\xi\rangle \\ &= \langle v\eta_{\mathcal{S}} | v\xi\rangle = \langle \eta_{\mathcal{S}} | \xi\rangle = \langle \eta_{\mathcal{S}} + \eta_{\mathcal{S}^\perp} | \xi\rangle = \langle \eta | \xi\rangle\,.\end{aligned}$$

In the third equality of the above computation we used the fact that v maps $\mathcal{S}^\perp$ to $\{0\}$ and in the fourth we used the property

$$\langle v\xi' | v\xi\rangle = \langle \xi' | \xi\rangle, \qquad \xi, \xi' \in \mathcal{S}$$

which follows from the fact that v is isometric on $\mathcal{S}$ via the polarization identity. □

A partial isometry v on $\mathcal{H}$ whose initial subspace is $\mathcal{H}$ (i.e. $v^*v = \mathbb{1}$) is called an *isometry*. This terminology is in agreement with the standard meaning of "isometry", as in this case v is an isometric mapping of $\mathcal{H}$ onto the subspace $v\mathcal{H} = vv^*\mathcal{H}$ of $\mathcal{H}$.

3.4 Polar Decomposition

Theorem 3.13
Let $x \in \mathrm{B}(\mathcal{H})$. Then there exists a unique pair of operators (v, a) such that
(1) $x = va$,
(2) *a is positive,*
(3) $v^*v = \mathbf{s}(a)$.

Proof
We begin by proving existence of a pair (v, a) satisfying conditions (1)–(3). Let $a = (x^*x)^{\frac{1}{2}}$. For any $\xi \in \mathcal{H}$ we have

$$\|a\xi\|^2 = \left\langle (x^*x)^{\frac{1}{2}}\xi \,\middle|\, (x^*x)^{\frac{1}{2}}\xi\right\rangle = \langle \xi \,|\, x^*x\xi\rangle = \langle x\xi \,|\, x\xi\rangle = \|x\xi\|^2.$$

It follows that there exists a linear map $v_{00} : a\mathcal{H} \to x\mathcal{H}$ such that for any $\xi \in \mathcal{H}$

$$v_{00}a\xi = x\xi.$$

(indeed: if $\eta \in a\mathcal{H}$, then there exists $\xi \in \mathcal{H}$ such that $\eta = a\xi$ and the vector $x\xi$ depends only on η because if $\eta = a\xi'$ for another $\xi' \in \mathcal{H}$ then $\|x\xi' - x\xi\| = \|x(\xi - \xi')\| = \|a(\xi - \xi')\| = \|\eta - \eta\| = 0$). It is clear that v_{00} is an isometric map, and so it extends uniquely to an isometry $v_0 : \overline{a\mathcal{H}} \to \mathcal{H}$. Now let v be the partial isometry defined as follows:

$$v\xi = \begin{cases} v_0\xi, & \xi \in \overline{a\mathcal{H}}, \\ 0, & \xi \in (a\mathcal{H})^\perp. \end{cases}$$

The operator v was defined precisely so that

$$va = x.$$

Moreover v^*v is the projection onto the initial subspace of v, i.e. onto the closure of the range of a. In other words

$$v^*v = \mathbf{l}(a) = \mathbf{s}(a),$$

because a is self-adjoint.

We now pass to the proof of uniqueness of the pair (v, a). Let (u, b) be a pair of operators such that

- $x = ub$,
- b is positive,
- $u^*u = \mathbf{s}(b)$.

Then

$$a^2 = x^*x = bu^*ub = b\mathbf{s}(b)b = b^2,$$

so by uniqueness of square roots we get $b = a$. Now, since u is a partial isometry with initial projection $\mathbf{s}(b) = \mathbf{s}(a)$, its initial subspace must be $\overline{a\mathcal{H}}$. Therefore

$$u\xi = 0, \qquad \xi \in (a\mathcal{H})^\perp.$$

On the other hand, for $\eta \in \mathcal{H}$ we have

$$u(a\eta) = x\eta = v_0(a\eta),$$

so u coincides with v_0 on $a\mathcal{H}$. It follows that $u = v$. □

The decomposition $x = va$ of $x \in \mathsf{B}(\mathcal{H})$ obtained in Theorem 3.13 is called the *polar decomposition* of x. The partial isometry v entering polar decomposition is sometimes referred to as the *phase* of x, while the positive operator a is called the *modulus* or *absolute value* of x and is denoted by the symbol $|x|$. Let us further note that if x is self-adjoint then $|x| = f(x)$, where $f(\lambda) = |\lambda|$ for all $\lambda \in \sigma(x)$.

Remark 3.14 Let $x \in \mathsf{B}(\mathcal{H})$ and let $x = v|x|$ be the polar decomposition of x.

(1) The phase v of x is unitary if and only if $\ker x = 0$ and the range of x is dense in $\mathcal{H}$.

(2) We have

$$x^* = |x|v^* = \mathbf{s}(|x|)|x|v^* = v^*v|x|v^* = v^*(v|x|v^*).$$

Moreover, the operator $v|x|v^*$ is positive and it is not hard to check that

$$\mathbf{s}(v|x|v^*) = v\mathbf{s}(|x|)v^* = vv^*vv^* = vv^*.$$

It follows that $x^* = v^*(v|x|v^*)$ is the polar decomposition of x^*.

(3) Assume that x is self-adjoint. Then $|x| = x^+ + x^-$ and $v = \mathbf{s}(x^+) - \mathbf{s}(x^-)$.

3.5 Monotone Convergence of Operators

In addition to metric topology defined by the operator norm there are a number of other useful topologies on $\mathsf{B}(\mathcal{H})$. On of them is the *strong topology* defined by the family of seminorms $\{p_\xi\}_{\xi \in \mathcal{H}}$, where

$$p_\xi(x) = \|x\xi\|, \qquad x \in \mathsf{B}(\mathcal{H}).$$

In this topology a net of operators $(x_i)_{i \in I}$ converges to $x \in \mathsf{B}(\mathcal{H})$ if and only if for any $\xi \in \mathcal{H}$ we have $x_i\xi \xrightarrow[i\in I]{} x\xi$, i.e. the operators x_i converge to x pointwise on $\mathcal{H}$.

Theorem 3.15
Let $(x_i)_{i \in I}$ be a net of self-adjoint operators such that

$$\bigl(i \succcurlyeq j\bigr) \Longrightarrow \bigl(x_i \geq x_j\bigr)$$

and assume that there exists a constant $C > 0$ such that $\|x_i\| \leq C$ for all $i \in I$. Then the net $(x_i)_{i \in I}$ has a supremum (least upper bound) in $\mathsf{B}(\mathcal{H})$, i.e. there exists a self-adjoint operator x such that $x \geq x_i$ for all $i \in I$ and any operator y such that $y \geq x_i$ for all $i \in I$ satisfies $y \geq x$. Moreover, the net $(x_i)_{i \in I}$ converges to x in strong topology.

Proof
We begin by noticing that for any $\zeta \in \mathcal{H}$ the net $\bigl(\langle\zeta|x_i\zeta\rangle\bigr)_{i \in I}$ is bounded and non-decreasing, so it converges to its supremum

$$\sup_{i \in I}\langle\zeta|x_i\zeta\rangle = \lim_{i \in I}\langle\zeta|x_i\zeta\rangle.$$

It follows that for each $\xi, \eta \in \mathcal{H}$ the net $\big(\langle \xi \,|\, x_i \eta\rangle\big)_{i\in I}$ converges, as

$$\langle \xi \,|\, x_i \eta\rangle = \tfrac{1}{4}\sum_{k=0}^{3} \mathrm{i}^k \left\langle \eta + \mathrm{i}^k \xi \,\middle|\, x_i(\eta + \mathrm{i}^k \xi)\right\rangle, \qquad i \in I.$$

We let

$$F(\xi, \eta) = \lim_{i\in I} \langle \xi \,|\, x_i \eta\rangle, \qquad \xi, \eta \in \mathcal{H}.$$

It is easy to see that F is a sesquilinear form which is bounded, since

$$\big|F(\xi, \eta)\big| = \big|\lim_{i\in I} \langle \xi \,|\, x_i \eta\rangle\big| = \lim_{i\in I} \big|\langle \xi \,|\, x_i \eta\rangle\big| \leq C\|\xi\|\|\eta\|, \qquad \xi, \eta \in \mathcal{H},$$

and therefore there exists a unique operator $x \in \mathsf{B}(\mathcal{H})$ such that

$$\langle \xi \,|\, x\eta\rangle = F(\xi, \eta) = \lim_{i\in I} \langle \xi \,|\, x_i \eta\rangle, \qquad \xi, \eta \in \mathcal{H}.$$

The operator x is self-adjoint, since $F(\xi, \xi)$ is real for any ξ (as a limit of real numbers, cf. Proposition 3.3(2)). Moreover $x = \sup_{i\in I} x_i$ in the sense of partial order on $\mathsf{B}(\mathcal{H})$. Indeed: for $\xi \in \mathcal{H}$ we have

$$\langle \xi \,|\, x\xi\rangle = \lim_{j\in I} \langle \xi \,|\, x_j \xi\rangle = \sup_{j\in I} \langle \xi \,|\, x_j \xi\rangle \geq \langle \xi \,|\, x_i \xi\rangle, \qquad i \in I$$

which means that $x \geq x_i$ for all $i \in I$. Furthermore, if $y \geq x_i$ for all i then for any $\xi \in \mathcal{H}$ we have

$$\langle \xi \,|\, y\xi\rangle \geq \langle \xi \,|\, x_i \xi\rangle, \qquad i \in I,$$

so that $\langle \xi \,|\, y\xi\rangle \geq \sup_{j\in I} \langle \xi \,|\, x_j \xi\rangle = \langle \xi \,|\, x\xi\rangle$. This means that $y \geq x$.

To see that $(x_i)_{i\in I}$ converges to x in strong topology take any $\xi \in \mathcal{H}$. We have

$$\begin{aligned}
\|x\xi - x_i\xi\|^2 &= \big\|(x - x_i)\xi\big\|^2 = \big\|(x - x_i)^{\frac{1}{2}}(x - x_i)^{\frac{1}{2}}\xi\big\|^2\\
&\leq \big\|(x - x_i)^{\frac{1}{2}}\big\|^2\big\|(x - x_i)^{\frac{1}{2}}\xi\big\|^2\\
&= \big\|(x - x_i)\big\|\big\|(x - x_i)^{\frac{1}{2}}\xi\big\|^2\\
&\leq 2C\big\|(x - x_i)^{\frac{1}{2}}\xi\big\|^2\\
&= 2C\left\langle (x - x_i)^{\frac{1}{2}}\xi \,\middle|\, (x - x_i)^{\frac{1}{2}}\xi\right\rangle\\
&= 2C\,\langle \xi \,|\, (x - x_i)\xi\rangle = 2C\big(F(\xi, \xi) - \langle \xi \,|\, x_i\xi\rangle\big) \xrightarrow[i\in I]{} 0,
\end{aligned}$$

so $x_i\xi \xrightarrow[i\in I]{} x\xi$. □

Notes

In this chapter we have laid the groundwork for further development of operator theory. A particularly crucial role in several of the following chapters will be played by polar decomposition of operators. Numerous exercises and examples devoted to topics mentioned above can be found in [Ped, Section 3.2], [ReSi_1, Chapters VI and VII].

Spectral Theorems and Functional Calculus

P. Sołtan, *A Primer on Hilbert Space Operators*, Compact Textbooks in Mathematics,
https://doi.org/10.1007/978-3-319-92061-0_4

One of the most important facts about operators on finite dimensional spaces equipped with a scalar product is the *spectral theorem* which says that a self-adjoint operator m can be written as a linear combination

$$m = \sum_{i=1}^{N} \lambda_i E_i \tag{4.1}$$

of pairwise orthogonal projections

$$\{E_1, \ldots, E_N\}$$

whose coefficients $\{\lambda_1, \ldots, \lambda_N\}$ are the (real) eigenvalues of m. As a consequence, for any polynomial $f \in \mathbb{C}[\,\cdot\,]$ we have

$$f(m) = \sum_{i=1}^{N} f(\lambda_i) E_i. \tag{4.2}$$

Indeed: since $E_1, \ldots, E_N$ are pairwise orthogonal, for $f(\lambda) = \lambda^n$ we obtain

$$f(m) = \left(\sum_{i=1}^{N} \lambda_i E_i \right)^n = \sum_{i_1, \ldots, i_n = 1}^{N} \lambda_{i_1} \ldots \lambda_{i_n} E_{i_1} \ldots E_{i_n} = \sum_{i=1}^{N} \lambda_i^m E_i$$

and the result for general f follows by linearity.

No doubt, in case of infinite dimensional Hilbert spaces, the spectral theorem needs to be appropriately reformulated. The search for an analog of the decomposition (4.1) leads to the notion of a *spectral measure* which will be discussed in ▶ Sect. 4.3. In this new formulation the linear combination (4.1) is replaced by an appropriately defined integral.

It turns out that some of the very useful forms of the spectral theorem generalize not so much the equality (4.1), as (4.2). In other words the spectral theorem in one of its forms deals with *functional calculus*. Functional calculus for continuous functions on the spectrum of a self-adjoint operator was the focus of ▶ Chap. 2. In this chapter we will broaden the class of functions for which functional calculus can be defined. As an intermediate step we will prove yet another formulation of the spectral theorem, namely we will show that a self-adjoint operator is unitarily equivalent to an operator of multiplication by a function on an L_2 space. Each of the above mentioned results can be referred to as the spectral theorem and each has important and interesting consequences. At the end of the chapter we will define holomorphic functional calculus and give a brief account of functional calculus in C*-algebras.

4.1 Multiplication Operators

Let (Ω, μ) be a *semifinite* measure space.[1] Any $f \in L_\infty(\Omega, \mu)$ defines a linear operator M_f on $L_2(\Omega, \mu)$:

$$(M_f\psi)(\omega) = f(\omega)\psi(\omega), \qquad \omega \in \Omega,\ \psi \in L_2(\Omega, \mu).$$

The operator M_f is called the *operator of multiplication* by f (*multiplication operator* for short). Clearly M_f is bounded and $\|M_f\| \leq \|f\|_\infty$.

Lemma 4.1 *For $f \in L_\infty(\Omega, \mu)$ we have*

$$\|f\|_\infty = \inf\left\{c > 0 \,\middle|\, |f| \dot{\leq} c\right\}.$$

Proof

By definition

$$\|f\|_\infty = \inf_{\mu(\Omega\setminus\Delta)=0}\ \sup_{\omega\in\Delta} \left|f(\omega)\right|.$$

Let $S = \left\{\sup\limits_{\omega\in\Delta} |f(\omega)| \,\middle|\, \mu(\Omega\setminus\Delta) = 0\right\}$ and $C = \left\{c > 0 \,\middle|\, |f| \dot{\leq} c\right\}$. We will show that $\inf C = \inf S$.

Let $s \in S$. Then there exists a measurable $\Delta_s \subset \Omega$ such that $\mu(\Omega\setminus\Delta_s) = 0$ and $s = \sup\limits_{\omega\in\Delta_s} \left|f(\omega)\right|$. Thus $|f| \leq s$ on Δ_s which is of full measure and therefore $s \in C$. It follows that $S \subset C$, and so $\inf S \geq \inf C$.

Let $c \in C$. Then there is a measurable $\Delta_c \subset \Omega$ with $\mu(\Omega\setminus\Delta_c) = 0$ and $|f| \leq c$ on Δ_c. Thus $s = \sup\limits_{\omega\in\Delta_c} \left|f(\omega)\right| \leq c$. In other words for any $c \in C$ there exists $s \in S$ such that $s \leq c$, which shows that $\inf C \geq \inf S$. □

[1] A measure space (Ω, μ) is *semifinite* if the measure μ is semifinite, i.e. for any measurable $\Delta \subset \Omega$ such that $\mu(\Delta) > 0$ there exists a measurable $\Lambda \subset \Delta$ such that $0 < \mu(\Lambda) + \infty$.

Let us note that $L_\infty(\Omega, \mu)$ is a C*-algebra with usual operations of addition and multiplication and involution $f \mapsto \overline{f}$.

Proposition 4.2 *The map*

$$L_\infty(\Omega, \mu) \ni f \longmapsto M_f \in \mathrm{B}\big(L_2(\Omega, \mu)\big)$$

is an isometric $*$-isomorphism onto its range.

Proof

The fact that $f \mapsto M_f$ preserves algebraic operations (including the involution) is rather obvious. We will check that the map in question is isometric. Let $\psi \in L_2(\Omega, \mu)$. For $c \geq \|f\|_\infty$ we have $c \dot{\geq} |f|$, so

$$\|f\psi\|_2^2 = \int_\Omega |f|^2|\psi|^2\, d\mu \leq c^2\|\psi\|_2^2.$$

Therefore $\|M_f\| \leq c$ and consequently $\|M_f\| \leq \|f\|_\infty$ by Lemma 4.1.

Conversely, if $0 < c < \|f\|_\infty$ then again by Lemma 4.1, the measure of the set $\big\{\omega \,\big|\, |f(\omega)| \geq c\big\}$ is non-zero. Thus there exists a measurable $\Lambda \subset \Omega$ such that $0 < \mu(\Lambda) < +\infty$ and $|f| \geq c$ on Λ. Let $\psi = \frac{1}{\sqrt{\mu(\Lambda)}}\chi_\Lambda$. Then $\|\psi\|_2 = 1$ and

$$\|f\psi\|_2^2 = \frac{1}{\mu(\Lambda)}\int_\Lambda |f|^2\, d\mu \geq c^2,$$

so that $\|M_f\| \geq c$. It follows that $\|M_f\| \geq \|f\|_\infty$. □

For a measurable function $f : \Omega \to \mathbb{C}$ we define the *essential range* of f as

$$V_{\text{ess}}(f) = \big\{\lambda \in \mathbb{C} \,\big|\, \text{for any neighborhood } \mathcal{U} \text{ of } \lambda \text{ we have } \mu\big(f^{-1}(\mathcal{U})\big) > 0\big\}.$$

Note that if $f \doteq g$ then $V_{\text{ess}}(f) = V_{\text{ess}}(g)$. Indeed: the condition that $f \doteq g$ means that $f = g$ on certain measurable subset $\Delta \subset \Omega$ such that $\mu(\Omega \setminus \Delta) = 0$. Therefore for any Borel subset $A \subset \mathbb{C}$

$$\begin{aligned}\mu\big(f^{-1}(A)\big) &= \mu\big(f^{-1}(A) \cap \Delta\big) = \mu\big(\{\omega \in \Delta \,|\, f(\omega) \in A\}\big)\\ &= \mu\big(\{\omega \in \Delta \,|\, g(\omega) \in A\}\big) = \mu\big(g^{-1}(A) \cap \Delta\big) = \mu\big(g^{-1}(A)\big).\end{aligned}$$

It is also easy to check that $V_{\text{ess}}(f)$ is a closed subset of $\mathbb{C}$.

Theorem 4.3

Let $f \in L_\infty(\Omega, \mu)$. Then $\sigma(M_f) = V_{\text{ess}}(f)$.

Proof

Take a $\lambda \in \mathbb{C} \setminus V_{\text{ess}}(f)$. Then there exists $r > 0$ such that the set $\{\omega \mid |\lambda - f(\omega)| < r\}$ has measure 0, i.e. $|\lambda - f(\omega)| \geq r$ almost everywhere on Ω. Therefore the function

$$g(\omega) = \tfrac{1}{\lambda - f(\omega)}, \qquad \omega \in \Omega$$

belongs to $L_\infty(\Omega, \mu)$ and it is clear that g is the inverse of $\lambda - f$. Thus the operator M_g is the inverse of $\lambda\mathbb{1} - M_f$.

Conversely, let $\lambda \in V_{\text{ess}}(f)$. Then for any n the set

$$\{\omega \mid |\lambda - f(\omega)| \leq \tfrac{1}{n}\}$$

has non-zero measure, and so there exists a measurable set Λ_n of finite and non-zero measure such that $|\lambda - f| \leq \frac{1}{n}$ on Λ_n. Let $\psi_n = \frac{1}{\sqrt{\mu(\Lambda_n)}}\chi_{\Lambda_n}$. We have $\|\psi_n\|_2 = 1$ for all n and

$$\big|\big(\lambda - f(\omega)\big)\psi_n(\omega)\big| \leq \tfrac{1}{n}\big|\psi_n(\omega)\big|, \qquad \omega \in \Omega,$$

so that $\|(\lambda\mathbb{1} - M_f)\psi_n\|_2 \leq \frac{1}{n}$. This, in turn, means that $\lambda\mathbb{1} - M_f$ is not invertible, since if y were the inverse of $\lambda\mathbb{1} - M_f$ then we would have

$$1 = \big\|y(\lambda\mathbb{1} - M_f)\psi_n\big\|_2 \leq \|y\|\big\|(\lambda\mathbb{1} - M_f)\psi_n\big\|_2 \xrightarrow[n\to\infty]{} 0.$$

□

At some point later on the following fact will be needed:

Proposition 4.4 *Let (Ω, μ) be a semifinite measure space and let $f \in L_\infty(\Omega, \mu)$. Then $\lambda \in \mathbb{C}$ is an eigenvalue of the operator $M_f \in \mathsf{B}\big(L_2(\Omega, \mu)\big)$ if and only if there exists a measurable $\Delta \subset \Omega$ such that $\mu(\Delta) > 0$ and $f \doteq \lambda$ on Δ.*

Proof

Assume there exists a set Δ with the properties listed in the formulation of the theorem. Then it contains a measurable Δ' such that $0 < \mu(\Delta') < +\infty$, and so $\psi = \frac{1}{\sqrt{\mu(\Delta')}}\chi_{\Delta'}$ is an element of $L_2(\Omega, \mu)$ such that $\|\psi\|_2 = 1$ and $M_f\psi = \lambda\psi$.

On the other hand, if there exists a non-zero $\psi \in L_2(\Omega, \mu)$ such that $M_f\psi = \lambda\psi$ then $f\psi \doteq \lambda\psi$. Let $\Delta = \{\omega \in \Omega \mid \psi(\omega) \neq 0\}$. Then we must have $f \doteq \lambda$ on Δ. Moreover $\mu(\Delta) > 0$, since $\psi \neq 0$. □

As is easily checked any operator M_f on $L_2(\Omega, \mu)$ is normal and thus we immediately have the following corollary:

Corollary 4.5 *We have $\|M_f\| = \sup\{|\lambda| \mid \lambda \in V_{ess}(f)\}$.*

Proof
$\|M_f\| = |\sigma(M_f)| = \sup\{|\lambda| \mid \lambda \in \sigma(M_f)\}$. □

Remark 4.6 Assume that M_f is self-adjoint (i.e. f is real-valued). Let g be a continuous function on $\sigma(M_f) = V_{\text{ess}}(f)$. Then $g(M_f) = M_{g\circ f}$. Indeed: the mapping

$$\mathrm{C}\big(\sigma(M_f)\big) \ni g \longmapsto M_{g\circ f} \in \mathrm{B}(\mathcal{H})$$

satisfies the conditions uniquely determining the continuous functional calculus for M_f (the fact that it is isometric follows from Corollary 4.5).

Theorem 4.7
Let $x \in \mathrm{B}(\mathcal{H})$ be a self-adjoint operator. Then there exists a semifinite measure space (Ω, μ), an essentially bounded measurable real-valued function F on Ω, and a unitary operator

$$u : L_2(\Omega, \mu) \longrightarrow \mathcal{H}$$

such that

$$x = uM_F u^*.$$

Proof
Let $\mathrm{A} = \big\{f(x) \mid f \in \mathrm{C}\big(\sigma(x)\big)\big\}$. For a vector $\xi \in \mathcal{H}$ we will denote by $\mathrm{A}\xi$ the subspace

$$\big\{f(x)\xi \mid f \in \mathrm{C}\big(\sigma(x)\big)\big\} \subset \mathcal{H}.$$

Let $\mathcal{R}$ be the family of orthonormal systems $\{\xi_i\}_{i\in I}$ in $\mathcal{H}$ such that $A\xi_i \perp A\xi_j$ for any $i, j \in I$ with $i \neq j$. We put a partial order on $\mathcal{R}$ via the relation of containment. By the Kuratowski-Zorn lemma there exists a maximal element $\{\xi_j\}_{j\in J}$ of $\mathcal{R}$.

Let us see that the subspace span $\left(\bigcup_{j\in J} \mathrm{A}\xi_j\right)$ is dense in $\mathcal{H}$. Indeed: if there were a vector ξ with $\|\xi\| = 1$ and $\xi \perp \mathrm{A}\xi_j$ for all j then for any $f, g \in \mathrm{C}\big(\sigma(x)\big)$

$$\big\langle f(x)\xi \,\big|\, g(x)\xi_j\big\rangle = \big\langle \xi \,\big|\, f(x)^* g(x)\xi_j\big\rangle = \big\langle \xi \,\big|\, (\overline{f}g)(x)\xi_j\big\rangle = 0, \qquad j \in J,$$

and so the system $\{\xi_j\}_{j\in J} \cup \{\xi\}$ would also belong to $\mathcal{R}$ which would contradict the maximality of $\{\xi_j\}_{j\in J}$.

Put $\Omega = J \times \sigma(x)$. By considering on J the discrete topology we obtain a locally compact topology on Ω (the product topology). Furthermore, let $\mathfrak{M}$ be the family of subsets $\Delta \subset \Omega$

such that for each j

$$\Delta_j = \{\lambda \in \sigma(x) \,|\, (j, \lambda) \in \Delta\}$$

is a Borel subset of $\sigma(x)$. Then $\mathfrak{M}$ is a σ-algebra.[2]

For any $j \in J$ the map

$$\mathrm{C}(\sigma(x)) \ni f \longmapsto \langle \xi_j | f(x)\xi_j \rangle \in \mathbb{C}$$

is a positive linear functional, so by the Riesz-Markov-Kakutani representation theorem there exists a positive regular Borel measure μ_j on $\sigma(x)$ such that

$$\langle \xi_j | f(x)\xi_j \rangle = \int_{\sigma(x)} f \, d\mu_j, \qquad f \in \mathrm{C}(\sigma(x)).$$

The measures $\{\mu_j\}_{j \in J}$ allow us to define a measure μ on $\mathfrak{M}$: for $\Delta \in \mathfrak{M}$ we put

$$\mu(\Delta) = \sum_{j \in J} \mu_j(\Delta_j).$$

It is easy to see that μ is a semifinite measure on Ω (all the measures μ_j are finite). Moreover

- $\psi \in L_2(\Omega, \mu)$ if and only if there exist $j_1, j_2, \ldots \in J$ such that $j \in J \setminus \{j_1, j_2, \ldots\}$ we have $\psi(j, \lambda) = 0$ for μ_j-almost all λ and

$$\sum_{n=1}^{\infty} \int_{\sigma(x)} |\psi(j_n, \lambda)|^2 d\mu_{j_n}(\lambda) < +\infty,$$

- the space $\mathrm{C_c}(\Omega)$, i.e. the space of continuous functions on Ω with compact support, is dense in $\in L_2(\Omega, \mu)$.

We will now define the operator $u \in \mathrm{B}(L_2(\Omega, \mu), \mathcal{H})$. On the dense subspace $\mathrm{C_c}(\Omega)$ we put

$$uf = \sum_{j \in J} f(j, x)\xi_j, \qquad f \in \mathrm{C_c}(\Omega)$$

[2]The fact that $\mathfrak{M}$ is a σ-algebra is rather obvious. Moreover, it turns out that $\mathfrak{M}$ is nothing else, but the σ-algebra of Borel subsets of Ω. Indeed: clearly any open subset of Ω belongs to $\mathfrak{M}$, so that $\mathfrak{M}$ contains all Borel sets. On the other hand if Δ is a Borel subset of Ω then for each $j \in J$ the set Δ_j can be identified with $\Delta \cap (\{j\} \times \sigma(x))$. This identification comes from the homeomorphism

$$\sigma(x) \ni \lambda \longmapsto (j, \lambda) \in \{j\} \times \sigma(x),$$

which, of course, preserves the Borel structure. It follows that each Δ_j is a Borel subset of $\sigma(x)$.

(the sum is finite, as the support of f is compact). Then for any $f \in C_c(\Omega)$ we have

$$\begin{aligned}\|uf\|^2 &= \Big\|\sum_{j\in J} f(j,x)\xi_j\Big\|^2 = \Big\langle \sum_{j\in J} f(j,x)\xi_j \Big| \sum_{i\in J} f(i,x)\xi_i \Big\rangle \\ &= \sum_{i,j\in J} \big\langle f(j,x)\xi_j \,|\, f(i,x)\xi_i\big\rangle = \sum_{j\in J}\big\langle f(j,x)\xi_j \,|\, f(j,x)\xi_j\big\rangle \\ &= \sum_{j\in J}\big\langle \xi_j \,|\, f(j,x)^* f(j,x)\xi_j\big\rangle = \sum_{j\in J}\Big\langle \xi_j \Big| |f|^2(j,x)\xi_j\Big\rangle \\ &= \sum_{j\in J}\int_{\sigma(x)} |f|^2(j,\lambda)\,d\mu_j(\lambda) = \int_\Omega |f|^2\,d\mu = \|f\|_2^2,\end{aligned}$$

and consequently u extends to an isometry $L_2(\Omega,\mu)\to\mathcal{H}$.

Note that the range of u contains $\bigcup_{j\in J} \mathsf{A}\xi_j$ which is linearly dense in $\mathcal{H}$. This implies that u is surjective and, hence, unitary.

Define $F:\Omega\to\mathbb{C}$ by

$$F(j,\lambda)=\lambda, \qquad (j,\lambda)\in\Omega.$$

Then F is bounded and continuous (so, in particular, measurable). Moreover, for any $f \in C_c(\Omega)\subset L_2(\Omega,\mu)$ we have

$$\begin{aligned}u(M_F f) &= u(Ff) = \sum_{j\in J}(Ff)(j,x)\xi_j \\ &= \sum_{j\in J} xf(j,x)\xi_j = x\sum_{j\in J} f(j,x)\xi_j = xuf.\end{aligned}$$

By continuity we obtain $uM_F\psi = xu\psi$ for all $\psi \in L_2(\Omega,\mu)$, which means that $uM_Fu^* = x$. □

Two operators $x \in \mathsf{B}(\mathcal{H})$ and $y \in \mathsf{B}(\mathcal{K})$ are *unitarily equivalent* if there exists a unitary $u \in \mathsf{B}(\mathcal{K},\mathcal{H})$ such that $x = uyu^*$. Theorem 4.7 says that any self-adjoint operator is unitarily equivalent to an operator of multiplication by a bounded function (even by one which is continuous on a locally compact topological space). It turns out that analogous statement holds also for any normal operator (Theorem 7.6). Of course, operators which are unitarily equivalent have identical spectra and share many other properties. Also it is sometimes easy to answer questions about a given self-adjoint operator using the fact that it is equivalent to a multiplication operator.

Remark 4.8

(1) Let $\mathsf{A}\subset\mathsf{B}(\mathcal{H})$ be a $*$-algebra of operators. A vector $\xi\in\mathcal{H}$ is *cyclic* for A if $\mathsf{A}\xi$ is a dense subspace of $\mathcal{H}$. Now if $x\in\mathsf{B}(\mathcal{H})$ is a self-adjoint operator such that $\mathsf{A} = C^*(x,\mathbb{1})$ has a cyclic vector in $\mathcal{H}$ then by the reasoning in the proof of Theorem 4.7 we

can show that x is unitarily equivalent to an operator of multiplication by the identity function on $\sigma(x)$.

(2) Let $x \in \mathsf{B}(\mathcal{H})$ and $y \in \mathsf{B}(\mathcal{K})$ be self-adjoint and unitarily equivalent via $u \in \mathsf{B}(\mathcal{K}, \mathcal{H})$, i.e. $x = uyu^*$. Then, as we already mentioned, we have $\sigma(x) = \sigma(y)$ and for any $f \in \mathsf{C}(\sigma(x))$ we have $f(x) = uf(y)u^*$. This follows immediately from the uniqueness of continuous functional calculus.

4.2 Borel Functional Calculus

Let $x \in \mathsf{B}(\mathcal{H})$ be a self-adjoint operator. In this section we will generalize the continuous functional calculus for x to a wider class of functions, namely the bounded Borel functions on $\sigma(x)$. We begin with a proposition on automatic continuity of certain $*$-homomorphisms.[3]

Proposition 4.9 *Let $\mathcal{K}$ be a Hilbert space and let B be a Banach $*$-algebra with unit. Consider a unital $*$-homomorphism $\Phi : \mathsf{B} \to \mathsf{B}(\mathcal{K})$. Then Φ is a contraction.*

Proof

Let $b \in \mathsf{B}$ and $\lambda \in \mathbb{C}$ be such that $\lambda\mathbb{1} - b$ is invertible in B. Then $\lambda\mathbb{1} - \Phi(b)$ is an invertible operator. In particular, putting b^*b instead of b we obtain

$$\sigma\big(\Phi(b^*b)\big) \subset \big\{\lambda \in \mathbb{C} \,\big|\, \lambda\mathbb{1} - b^*b \text{ is not invertible in } \mathsf{B}\big\}$$

Therefore

$$\begin{aligned}\big\|\Phi(b)\big\|^2 &= \big\|\Phi(b)^*\Phi(b)\big\| = \big\|\Phi(b^*b)\big\| = \big|\sigma\big(\Phi(b^*b)\big)\big| \\ &\le \sup\big\{|\lambda| \,\big|\, \lambda\mathbb{1} - b^*b \text{ is not invertible in } \mathsf{B}\big\} \le \|b^*b\| = \|b\|^2,\end{aligned}$$

where the last equality follows from the fact that if $|\lambda| > \|b^*b\|$ then $\lambda\mathbb{1} - b^*b$ is invertible with

$$(\lambda\mathbb{1} - b^*b)^{-1} = \sum_{n=0}^{\infty} \lambda^{-n-1}(b^*b)^n$$

(the series converges in the Banach algebra B). This argument shows that every unital $*$-homomorphism from a Banach $*$-algebra to $\mathsf{B}(\mathcal{K})$ is contractive. □

Note that if X is a topological space then the set $\mathscr{B}(X)$ of bounded Borel functions on X is a Banach $*$-algebra (in fact, a C*-algebra) with natural operations of addition and multiplication, involution $f \mapsto \overline{f}$ and the uniform norm $\|\cdot\|_\infty$.

[3] A $*$-*homomorphism* is a linear, multiplicative and $*$-preserving map. A homomorphism $\Phi : \mathsf{A} \to \mathsf{B}$ between unital algebras is called *unital* when $\Phi(\mathbb{1}_{\mathsf{A}}) = \mathbb{1}_{\mathsf{B}}$.

The extension of functional calculus for self-adjoint operators from the class of continuous functions to bounded Borel functions will be based on Theorem 4.7 and the observation made in Remark 4.6.

Theorem 4.10 (Borel Functional Calculus)
Let $x \in \mathsf{B}(\mathcal{H})$ be a self-adjoint operator and denote by $\mathscr{B}(\sigma(x))$ the $*$-algebra of bounded Borel functions on $\sigma(x)$. Then there exists a unique unital $*$-homomorphism $\mathscr{B}(\sigma(x)) \to \mathsf{B}(\mathcal{H})$ denoted by

$$\mathscr{B}(\sigma(x)) \ni f \longmapsto f(x) \in \mathsf{B}(\mathcal{H})$$

such that
- *if $f(\lambda) = \lambda$ for all $\lambda \in \sigma(x)$ then $f(x) = x$,*
- *if $(f_n)_{n\in\mathbb{N}}$ is a uniformly bounded sequence of Borel functions converging pointwise to f then $f_n(x) \xrightarrow[n\to\infty]{} f(x)$ in strong topology.*

Moreover this homomorphism extends the isomorphism $\mathsf{C}(\sigma(x)) \to \mathsf{C}^*(x, \mathbb{1})$ given by the continuous functional calculus.

Proof
The operator x is unitarily equivalent to an operator of multiplication by a Borel (in fact even continuous) function F on a Hilbert space $L_2(\Omega, \mu)$. Moreover the range of F is $\sigma(x)$, so for any bounded Borel function f on $\sigma(x)$ the composition $f \circ F$ is measurable (with respect to the Borel structure on Ω) and bounded. Define

$$f(x) = uM_{f\circ F}u^*, \qquad f \in \mathscr{B}(\sigma(x)).$$

It is clear that $f \mapsto f(x)$ is a unital $*$-homomorphism and if $f(\lambda) = \lambda$ for all $\lambda \in \sigma(x)$ then $f(x) = x$. Incidentally, it is also clear that $f \mapsto f(x)$ is a contraction, as $\|f(x)\| = \|uM_{f\circ F}u^*\| = \|M_{f\circ F}\| \leq \|f\|_\infty$, so in this case we do not need to use Proposition 4.9.

Let $(f_n)_{n\in\mathbb{N}}$ be a bounded sequence of Borel functions on $\sigma(x)$ converging pointwise to f and take $\xi \in \mathcal{H}$. Setting $\psi = u^*\xi$ and using the fact that u is isometric we obtain

$$\begin{aligned}\|f_n(x)\xi - f(x)\xi\|^2 &= \|uM_{(f_n-f)\circ F}u^*\xi\|^2 \\ &= \|uM_{(f_n-f)\circ F}\psi\|^2 \\ &= \|M_{(f_n-f)\circ F}\psi\|^2 \\ &= \int_\Omega |f_n(F(\omega)) - f(F(\omega))|^2 |\psi(\omega)|^2 \, d\mu(\omega) \xrightarrow[n\to\infty]{} 0\end{aligned}$$

by the dominated convergence theorem.

We have already established that the map $\mathscr{B}\big(\sigma(x)\big) \ni f \mapsto f(x) \in \mathsf{B}(\mathcal{H})$ is a contractive unital $*$-homomorphism and that it maps the identity function to x. It is therefore immediate from the uniqueness of continuous functional calculus that it extends the latter.

Let us address the uniqueness of the homomorphism

$$\mathscr{B}\big(\sigma(x)\big) \ni f \longmapsto f(x) \in \mathsf{B}(\mathcal{H}).$$

Let $\Phi : \mathscr{B}\big(\sigma(x)\big) \to \mathsf{B}(\mathcal{H})$ be a unital $*$-homomorphism (which by Proposition 4.9 is necessarily contractive) mapping the identity function to x and such that if $(f_n)_{n\in\mathbb{N}}$ is a bounded sequence of Borel functions converging pointwise to f then $\Phi(f_n)$ converges to $\Phi(f)$ in strong topology. It follows from the uniqueness of continuous functional calculus that if f is continuous then $\Phi(f) = f(x)$.

For an open set $\Delta \subset \sigma(x)$ choose a bounded sequence of continuous functions $(f_n)_{n\in\mathbb{N}}$ converging pointwise to χ_Δ. Then for any $\xi \in \mathcal{H}$ we have

$$\chi_\Delta(x)\xi = \lim_{n\to\infty} f_n(x)\xi = \lim_{n\to\infty} \Phi(f_n)\xi = \Phi(\chi_\Delta)\xi.$$

Thus the family

$$\mathcal{L} = \big\{\Delta \in \mathfrak{M} \,\big|\, \Phi(\chi_\Delta) = \chi_\Delta(x)\big\}$$

contains all open sets. This family is also a $\boldsymbol{\lambda}$-system (see Appendix A.2):

(1) Ω is open, so $\Omega \in \mathcal{L}$,

(2) if $\Delta \in \mathcal{L}$ then

$$\Phi(\chi_{\Delta^{\mathsf{C}}}) = \Phi(\chi_\Omega - \chi_\Delta) = \mathbb{1} - \Phi(\chi_\Delta) = \mathbb{1} - \chi_\Delta(x) = \chi_{\Delta^{\mathsf{C}}}(x)$$

which means that $\Delta^{\mathsf{C}} \in \mathcal{L}$,

(3) if $(\Delta_n)_{n\in\mathbb{N}}$ is a sequence of pairwise disjoint elements of $\mathcal{L}$ and $\Delta = \bigcup\limits_{n=1}^{\infty} \Delta_n$ then

$$\chi_\Delta = \sum_{n=1}^{\infty} \chi_{\Delta_n} = \lim_{N\to\infty} \sum_{n=1}^{N} \chi_{\Delta_n},$$

i.e. χ_Δ is a pointwise limit of the sequence $\bigg(\sum\limits_{n=1}^{N} \chi_{\Delta_n}\bigg)_{N\in\mathbb{N}}$ and for any $\xi \in \mathcal{H}$

$$\Phi(\chi_\Delta)\xi = \lim_{N\to\infty} \sum_{n=1}^{N} \Phi(\chi_{\Delta_n})\xi = \lim_{N\to\infty} \sum_{n=1}^{N} \chi_{\Delta_n}(x)\xi = \chi_\Delta(x)\xi.$$

Therefore $\Delta \in \mathcal{L}$.

Since the family of all open subsets of $\sigma(x)$ is a $\boldsymbol{\pi}$-system, by Dynkin's theorem on $\boldsymbol{\pi}$- and $\boldsymbol{\lambda}$-systems (Theorem A.2), we have $\mathfrak{M} \subset \mathcal{L}$. Now any bounded Borel function is a pointwise limit of a bounded sequence of simple functions, so the homomorphisms Φ and $f \mapsto f(x)$ agree on all of $\mathscr{B}\big(\sigma(x)\big)$. □

4.3 Spectral Measures

Let Ω be a set and let $\mathfrak{M}$ be a σ-algebra of subsets of Ω. A *spectral measure* on Ω is a map $E : \mathfrak{M} \to \operatorname{Proj}\big(\mathsf{B}(\mathcal{H})\big)$ such that

- $E(\emptyset) = 0$, $E(\Omega) = \mathbb{1}$,
- for any $\Delta_1, \Delta_2 \in \mathfrak{M}$ we have $E(\Delta_1 \cap \Delta_2) = E(\Delta_1)E(\Delta_2)$,
- for pairwise disjoint $\Delta_1, \Delta_2, \ldots \in \mathfrak{M}$ we have $E\left(\bigcup\limits_{n=1}^{\infty} \Delta_n\right) = \sum\limits_{n=1}^{\infty} E(\Delta_n)$.

The sum in the last condition is taken to mean the limit of finite sums in strong topology. Note that it follows from the first two conditions that projections corresponding to disjoint sets are orthogonal and hence their sum is a projection.

Fundamental examples of spectral measures arise from self-adjoint operators: let $x \in \mathsf{B}(\mathcal{H})$ be self-adjoint, put $\Omega = \sigma(x)$ and let $\mathfrak{M}$ be the σ-algebra of Borel subsets of $\sigma(x)$. Define

$$E_x : \mathfrak{M} \ni \Delta \longmapsto \chi_\Delta(x) \in \operatorname{Proj}\big(\mathsf{B}(\mathcal{H})\big). \tag{4.3}$$

Then E_x is a spectral measure. We will see later on that this measure determines x uniquely.

Let E be a spectral measure on Ω. Then for any $\xi \in \mathcal{H}$ the formula

$$\Delta \longmapsto \langle \xi \,|\, E(\Delta)\xi \rangle, \qquad \Delta \in \mathfrak{M}$$

defines a finite (positive) measure on Ω. We will denote this measure by the symbol $\langle \xi \,|\, E\xi \rangle$, so that the integral of a function f with respect to this measure will be written as

$$\int\limits_{\Omega} f \, d\langle \xi \mid E\xi \rangle \quad \text{or} \quad \int\limits_{\Omega} f(\omega) \, d\langle \xi \mid E(\omega)\xi \rangle.$$

Similarly, for $\xi, \eta \in \mathcal{H}$ the mapping

$$\langle \xi \,|\, E\eta \rangle : \Delta \longmapsto \langle \xi \,|\, E(\Delta)\eta \rangle, \qquad \Delta \in \mathfrak{M}$$

is a complex measure with finite *total variation*. Let us recall that the total variation of a measure ν is the supremum of all sums of the form $\sum\limits_{n=1}^{N} \big|\nu(\Delta_n)\big|$, where $\{\Delta_1, \ldots, \Delta_N\}$ is a finite partition of Ω into measurable sets. for $\nu = \langle \xi \,|\, E\eta \rangle$ the total variation is bounded by $\|\xi\|\|\eta\|$. Indeed: for any partition $\{\Delta_1, \ldots, \Delta_N\}$ there are complex numbers

$\lambda_1, \ldots, \lambda_N$ of modulus one such that

$$\sum_{n=1}^{N} \left| \langle \xi \,|\, E\eta \rangle (\Delta_n) \right| = \sum_{n=1}^{N} \left| \langle \xi \,|\, E(\Delta_n)\eta \rangle \right| = \sum_{n=1}^{N} \lambda_n \langle \xi \,|\, E(\Delta_n)\eta \rangle = \langle \xi \,|\, t\eta \rangle ,$$

where $t = \sum_{n=1}^{N} \lambda_n E(\Delta_n)$. Since the projections $E(\Delta_1), \ldots, E(\Delta_N)$ are pairwise orthogonal, we have

$$\begin{aligned} \|t\|^2 = \|t^* t\| &= \left\| \left(\sum_{n=1}^{N} \overline{\lambda_n} E(\Delta_n) \right) \left(\sum_{m=1}^{N} \lambda_m E(\Delta_m) \right) \right\| \\ &= \left\| \sum_{m,n=1}^{N} \overline{\lambda_n} \lambda_m E(\Delta_n) E(\Delta_m) \right\| = \left\| \sum_{n=1}^{N} E(\Delta_n) \right\| = \|\mathbb{1}\| = 1 \end{aligned}$$

which shows that the total variation of $\langle \xi \,|\, E\eta \rangle$ is not greater than $\|\xi\| \|\eta\|$.

Let us notice further that for any bounded measurable function f on Ω the quantity

$$\int_{\Omega} f(\omega)\, d\langle \xi \mid E(\omega)\eta \rangle$$

is a sesquilinear form with respect to the variables (ξ, η) which is bounded by $\|\xi\| \|\eta\| \|f\|_\infty$. Consequently it defines a bounded operator $x_f \in \mathsf{B}(\mathcal{H})$ via

$$\langle \xi \,|\, x_f \eta \rangle = \int_{\Omega} f(\omega)\, d\langle \xi \mid E(\omega)\eta \rangle, \qquad \xi, \eta \in \mathcal{H}.$$

Theorem 4.11

The map $f \mapsto x_f \in \mathsf{B}(\mathcal{H})$ is a $*$-homomorphism from the algebra of bounded measurable functions on Ω into $\mathsf{B}(\mathcal{H})$. Moreover, if $(f_n)_{n \in \mathbb{N}}$ is a uniformly bounded sequence of functions converging pointwise to f then

$$x_{f_n} \xrightarrow[n \to \infty]{} x_f$$

in strong topology.

Proof
First let us notice that the map $f \mapsto x_f$ is a contraction for the norm $\|\cdot\|_\infty$ on bounded measurable functions and the operator norm, since

$$\|x_f\| = \sup_{\|\xi\|=\|\eta\|=1} \left|\left\langle \xi \middle| x_f \eta\right\rangle\right| = \sup_{\|\xi\|=\|\eta\|=1} \left| \int_\Omega f \, d\langle \xi \mid E\eta\rangle \right|$$
$$\leq \sup_\Omega |f| \cdot \big(\text{total variation of } \langle \xi \mid E\eta\rangle\big) \leq \sup_\Omega |f|.$$

Therefore it will be enough to prove multiplicativity

$$x_f x_g = x_{fg}$$

for simple functions only, as they are uniformly dense in the set of bounded measurable functions. Let

$$f = \sum_{n=1}^N \alpha_n \chi_{\Delta_n}, \quad g = \sum_{m=1}^M \beta_m \chi_{\Lambda_m}.$$

We have

$$x_f x_g = \left(\sum_{n=1}^N \alpha_n E(\Delta_n)\right)\left(\sum_{m=1}^M \beta_m E(\Lambda_m)\right)$$
$$= \sum_{m,n} \alpha_n \beta_m E(\Delta_n) E(\Lambda_m)$$
$$= \sum_{m,n} \alpha_n \beta_m E(\Delta_n \cap \Lambda_m) = x_{fg}.$$

Analogous argument shows that $x_{\overline{f}} = x_f{}^*$.

Now let $(f_n)_{n\in\mathbb{N}}$ be a bounded sequence of measurable functions converging pointwise to f. Then for any $\xi \in \mathcal{H}$

$$\|x_{f_n}\xi - x_f\xi\|^2 = \left\langle \xi \middle| x_{|f_n - f|^2}\xi\right\rangle = \int_\Omega \left|f_n(\omega) - f(\omega)\right|^2 d\langle \xi \mid E(\omega)\xi\rangle \xrightarrow[n\to\infty]{} 0$$

by the dominated converge theorem. □

Let us introduce the following notation for operators x_f:

$$x_f = \int_\Omega f(\omega)\, dE(\omega). \tag{4.4}$$

Let $x \in \mathsf{B}(\mathcal{H})$ be self-adjoint and let E_x be the spectral measure on $\sigma(x)$ defined by (4.3). Then it is easy to check that $f \mapsto f(x)$ and $f \mapsto x_f$ agree on simple functions:

for $f = \sum_{n=1}^{N} \lambda_n \chi_{\Delta_n}$ both $f(x)$ and x_f are equal to

$$\sum_{n=1}^{N} \lambda_n E_x(\Delta_n).$$

Since any Borel function is a pointwise limit of simple functions, we get $f(x) = x_f$ for all $f \in \mathscr{B}\big(\sigma(x)\big)$, i.e.

$$f(x) = \int_{\sigma(x)} f(\lambda)\, dE_x(\lambda), \qquad f \in \mathscr{B}\big(\sigma(x)\big).$$

In particular for $f(\lambda) = \lambda$ we have

$$x = \int_{\sigma(x)} \lambda\, dE_x(\lambda). \tag{4.5}$$

For a fixed self-adjoint operator $x \in \mathsf{B}(\mathcal{H})$ a spectral measure E_x such that (4.5) holds is unique. To see that note that if

$$x = \int_{\sigma(x)} \lambda\, dE(\lambda)$$

for some Borel spectral measure E on $\sigma(x)$ then the mapping

$$\mathscr{B}\big(\sigma(x)\big) \ni f \longmapsto \int_{\sigma(x)} f(\lambda)\, dE(\lambda)$$

satisfies the conditions uniquely defining the Borel functional calculus for x. Therefore for each bounded Borel function f we have

$$\int_{\sigma(x)} f(\lambda)\, dE(\lambda) = f(x) = \int_{\sigma(x)} f(\lambda)\, dE_x(\lambda).$$

Applying this to $f = \chi_\Delta$ for an arbitrary Borel set $\Delta \subset \sigma(x)$ we obtain

$$E(\Delta) = E_x(\Delta),$$

so the two measures are equal.

4.4 Holomorphic Functional Calculus

Let $x \in \mathsf{B}(\mathcal{H})$ be an arbitrary operator. In particular, x is not assumed to be self-adjoint. The algebra of functions holomorphic on a neighborhood of $\sigma(x)$ will be denoted by the symbol $\mathscr{H}\big(\sigma(x)\big)$.[4]

Let $f \in \mathscr{H}\big(\sigma(x)\big)$ and let Γ be a positively oriented curve surrounding $\sigma(x)$ contained in the domain of holomorphy of f. Define

$$f(x) = \tfrac{1}{2\pi\mathrm{i}} \oint\limits_{\Gamma} f(\lambda)(\lambda\mathbb{1} - x)^{-1} d\lambda$$

(the integral is of a continuous Banach space-valued function over a compact set). As usual, the value of the integral is independent of the choice of the curve Γ. Moreover, it is easy to see that $f(x)$ commutes with x.

Theorem 4.12 (Holomorphic Functional Calculus)
Let $x \in \mathsf{B}(\mathcal{H})$. Then
(1) *if f is a polynomial function $f(\lambda) = \alpha_0 + \alpha_1\lambda + \cdots + \alpha_n\lambda^n$ then*

$$f(x) = \alpha_0\mathbb{1} + \alpha_1 x + \cdots + \alpha_n x^n,$$

(2) *the map $\mathscr{H}\big(\sigma(x)\big) \ni f \mapsto f(x) \in \mathsf{B}(\mathcal{H})$ is a unital homomorphism.*

Proof
Let us first consider the case of $f(\lambda) = \lambda^m$ for some $m \in \mathbb{Z}_+$. Let Γ be a positively oriented circle around 0 with radius $r > \|x\|$. For $\lambda \in \Gamma$ we have

$$(\lambda\mathbb{1} - x)^{-1} = \sum_{n=0}^{\infty} \lambda^{-n-1} x^n.$$

Therefore

$$\begin{aligned} f(x) &= \tfrac{1}{2\pi\mathrm{i}} \oint\limits_{\Gamma} f(\lambda)(\lambda\mathbb{1} - x)^{-1} d\lambda \\ &= \sum_{n=0}^{\infty} \tfrac{1}{2\pi\mathrm{i}} \oint\limits_{\Gamma} \lambda^{m-n-1} x^n \, d\lambda \\ &= \sum_{n=0}^{\infty} \Bigg(\tfrac{1}{2\pi\mathrm{i}} \oint\limits_{\Gamma} \lambda^{m-n-1} \, d\lambda \Bigg) x^n = x^m. \end{aligned}$$

[4]More precisely, elements of $\mathscr{H}\big(\sigma(x)\big)$ are equivalence classes of the equivalence relation identifying functions which coincide on some neighborhood of $\sigma(x)$.

As the map $f \mapsto f(x)$ clearly is linear, we immediately infer that if f is a polynomial $f(\lambda) = \alpha_0 + \alpha_1\lambda + \cdots + \alpha_n\lambda^n$ then

$$f(x) = \alpha_0 \mathbb{1} + \alpha_1 x + \cdots + \alpha_n x^n.$$

Now we will show that the map $f \mapsto f(x)$ is multiplicative. Let $f, g \in \mathscr{H}(\sigma(x))$ and let Γ and Γ' be curves around $\sigma(x)$ contained within the intersection of domains of holomorphy of f and g, and such that Γ' lies outside of Γ. Using the resolvent identity

$$(\lambda\mathbb{1} - x)^{-1} - (\mu\mathbb{1} - x)^{-1} = (\mu - \lambda)(\lambda\mathbb{1} - x)^{-1}(\mu\mathbb{1} - x)^{-1}$$

(see Remark 1.3) we compute:

$$\begin{aligned}
f(x)g(x) &= \left(\tfrac{1}{2\pi i}\right)^2 \left(\oint_\Gamma f(\lambda)(\lambda\mathbb{1} - x)^{-1} d\lambda \right)\left(\oint_{\Gamma'} g(\mu)(\mu\mathbb{1} - x)^{-1} d\mu \right) \\
&= \left(\tfrac{1}{2\pi i}\right)^2 \oint_\Gamma \oint_{\Gamma'} f(\lambda)g(\mu)(\lambda\mathbb{1} - x)^{-1}(\mu\mathbb{1} - x)^{-1} d\mu\, d\lambda \\
&= \left(\tfrac{1}{2\pi i}\right)^2 \oint_\Gamma \oint_{\Gamma'} \tfrac{f(\lambda)g(\mu)}{\mu-\lambda} \left((\lambda\mathbb{1} - x)^{-1} - (\mu\mathbb{1} - x)^{-1}\right) d\mu\, d\lambda \\
&= \left(\tfrac{1}{2\pi i}\right)^2 \oint_\Gamma \oint_{\Gamma'} \tfrac{f(\lambda)g(\mu)}{\mu-\lambda} (\lambda\mathbb{1} - x)^{-1} d\mu\, d\lambda \\
&\qquad - \left(\tfrac{1}{2\pi i}\right)^2 \oint_\Gamma \oint_{\Gamma'} \tfrac{f(\lambda)g(\mu)}{\mu-\lambda} (\mu\mathbb{1} - x)^{-1} d\mu\, d\lambda \\
&= \left(\tfrac{1}{2\pi i}\right)^2 \oint_\Gamma f(\lambda)(\lambda\mathbb{1} - x)^{-1} \left(\oint_{\Gamma'} \tfrac{g(\mu)}{\mu-\lambda}\, d\mu \right) d\lambda \\
&\qquad - \left(\tfrac{1}{2\pi i}\right)^2 \oint_{\Gamma'} g(\mu)(\mu\mathbb{1} - x)^{-1} \underbrace{\left(\oint_\Gamma \tfrac{f(\lambda)}{\mu-\lambda} d\lambda \right)}_{=0} d\mu \\
&= \left(\tfrac{1}{2\pi i}\right)^2 \oint_\Gamma f(\lambda)(\lambda\mathbb{1} - x)^{-1} \left(\oint_{\Gamma'} \tfrac{g(\mu)}{\mu-\lambda}\, d\mu \right) d\lambda \\
&= \tfrac{1}{2\pi i} \oint_\Gamma f(\lambda) \left(\tfrac{1}{2\pi i} \oint_{\Gamma'} \tfrac{g(\mu)}{\mu-\lambda}\, d\mu \right) (\lambda\mathbb{1} - x)^{-1} d\lambda \\
&= \tfrac{1}{2\pi i} \oint_\Gamma f(\lambda)g(\lambda)(\lambda\mathbb{1} - x)^{-1} d\lambda = (fg)(x).
\end{aligned}$$

The indicated integral is equal to 0 because $\mu \in \Gamma'$ and Γ lies inside Γ', so we are integrating a holomorphic function over a curve contractible within its domain of holomorphy. □

Proposition 4.13 *Let $x \in \mathrm{B}(\mathcal{H})$. Then for any $f \in \mathscr{H}\big(\sigma(x)\big)$ we have $\sigma\big(f(x)\big) = f\big(\sigma(x)\big)$.*

Proof
Take $\mu \in \mathbb{C}\setminus f\big(\sigma(x)\big)$. Then the function $h_\mu : \lambda \mapsto \big(\mu - f(\lambda)\big)^{-1}$ is holomorphic on a neighborhood of $\sigma(x)$ and we have

$$h_\mu(x)\big(\mu\mathbb{1} - f(x)\big) = \big(\mu\mathbb{1} - f(x)\big)h_\mu(x) = \mathbb{1},$$

so that $\mu \notin \sigma\big(f(x)\big)$. In other words $\sigma\big(f(x)\big) \subset f\big(\sigma(x)\big)$.

Now let $\mu \in f\big(\sigma(x)\big)$. Then there exists $\lambda_0 \in \sigma(x)$ such that $\mu = f(\lambda_0)$ and

$$\mu - f(\lambda) = f(\lambda_0) - f(\lambda) = (\lambda_0 - \lambda)\tfrac{f(\lambda_0)-f(\lambda)}{\lambda_0-\lambda} = (\lambda_0 - \lambda)g(\lambda), \tag{4.6}$$

where

$$g(\lambda) = \begin{cases} \tfrac{f(\lambda_0)-f(\lambda)}{\lambda_0-\lambda}, & \lambda \neq \lambda_0, \\ f'(\lambda_0), & \lambda = \lambda_0 \end{cases}$$

belongs to $\mathscr{H}\big(\sigma(x)\big)$. Applying both sides of (4.6) to x we obtain

$$\mu\mathbb{1} - f(x) = (\lambda_0\mathbb{1} - x)g(x).$$

If the operator $\mu\mathbb{1} - f(x)$ were invertible, so would have to be $\lambda_0\mathbb{1} - x$, as $\lambda_0\mathbb{1} - x$ and $g(x)$ commute. However, $\lambda_0 \in \sigma(x)$, so it follows that $\mu \in \sigma\big(f(x)\big)$. This shows that $f\big(\sigma(x)\big) \subset \sigma\big(f(x)\big)$. □

Proposition 4.14 *Let $x \in \mathrm{B}(\mathcal{H})$, $g \in \mathscr{H}\big(\sigma(x)\big)$ and $f \in \mathscr{H}\Big(\sigma\big(g(x)\big)\Big)$. Then*

$$f\big(g(x)\big) = (f \circ g)(x)$$

Proof
Let Γ be a positively oriented curve surrounding $\sigma(x)$ within domain of holomorphy of g and let Γ' be a positively oriented curve in the domain of holomorphy of f surrounding $\sigma\big(g(x)\big) = g\big(\sigma(x)\big)$ and lying outside the image of Γ under g. For $\mu \in \Gamma'$ let $h_\mu(\lambda) = \frac{1}{\mu - g(\lambda)}$. Then h_μ is holomorphic on a neighborhood of $\sigma(x)$ and $h_\mu(x) = \big(\mu\mathbb{1} - g(x)\big)^{-1}$.

Thus

$$\begin{aligned} f\big(g(x)\big) &= \tfrac{1}{2\pi\mathrm{i}}\oint_{\Gamma'} f(\mu)\big(\mu\mathbb{1} - g(x)\big)^{-1}d\mu \\ &= \tfrac{1}{2\pi\mathrm{i}}\oint_{\Gamma'} f(\mu)h_\mu(x)\,d\mu \end{aligned}$$

$$
\begin{aligned}
&= \frac{1}{2\pi i}\oint_{\Gamma'} f(\mu)\left(\frac{1}{2\pi i}\oint_{\Gamma}\frac{1}{\mu-g(\lambda)}(\lambda\mathbb{1}-x)^{-1}d\lambda\right)d\mu \\
&= \frac{1}{2\pi i}\oint_{\Gamma}\left(\frac{1}{2\pi i}\oint_{\Gamma'}\frac{f(\mu)}{\mu-g(\lambda)}\,d\mu\right)(\lambda\mathbb{1}-x)^{-1}d\lambda \\
&= \frac{1}{2\pi i}\oint_{\Gamma}(f\circ g)(\lambda)(\lambda\mathbb{1}-x)^{-1}d\lambda = (f\circ g)(x).
\end{aligned}
$$

□

Let $D \subset \mathbb{C}$ be an open set and denote by $\overline{D}$ the set $\{\overline{z}\,|\,z \in D\}$. Let f be a holomorphic function on D. Then we can define a function $\widetilde{f} : \overline{D} \to \mathbb{C}$ putting

$$\widetilde{f}(z) = \overline{f(\overline{z})}, \qquad z \in \overline{D}.$$

It is easy to check that $\widetilde{f}$ is holomorphic on $\overline{D}$.

Proposition 4.15 *Let $x \in \mathrm{B}(\mathcal{H})$ and $f \in \mathscr{H}\big(\sigma(x)\big)$. Then $f(x)^* = \widetilde{f}(x^*)$.*

Proof
Choose a positively oriented curve Γ surrounding $\sigma(x)$ within domain of holomorphy of f. Denote by $\overline{\Gamma}'$ the image of Γ under complex conjugation with negative orientation, and by $\overline{\Gamma}$ the same curve with positive orientation we obtain

$$
\begin{aligned}
f(x)^* &= \left(\frac{1}{2\pi i}\oint_{\Gamma} f(\lambda)(\lambda\mathbb{1}-x)^{-1}d\lambda\right)^* \\
&= \frac{-1}{2\pi i}\oint_{\Gamma}\overline{f(\lambda)}(\overline{\lambda}\mathbb{1}-x^*)^{-1}d\overline{\lambda} \\
&= \frac{-1}{2\pi i}\left(-\oint_{\overline{\Gamma}'}\overline{f(\overline{\mu})}(\mu\mathbb{1}-x^*)^{-1}d\mu\right) \\
&= \frac{1}{2\pi i}\oint_{\overline{\Gamma}}\widetilde{f}(\mu)(\mu\mathbb{1}-x^*)^{-1}d\mu.
\end{aligned}
$$

□

Proposition 4.16 *Let $x \in \mathrm{B}(\mathcal{H})$ be self-adjoint. Then the holomorphic functional calculus for x is the restriction to $\mathscr{H}\big(\sigma(x)\big)$ of the isomorphism*

$$\mathrm{C}\big(\sigma(x)\big) \longrightarrow \mathrm{C}^*(x, \mathbb{1})$$

given by the continuous functional calculus.

Proof
For a temporary distinction, denote the image of $f \in \mathscr{H}(\sigma(x))$ under holomorphic functional calculus by $\Phi(f)$ instead of $f(x)$, reserving the latter symbol for the image of f under continuous functional calculus. For $f \in \mathscr{H}(\sigma(x))$ the operator $\Phi(f)$ is normal (by Proposition 4.15), and so

$$\|\Phi(f)\| = |\sigma(\Phi(f))| = \sup\{|f(\lambda)| \,|\, \lambda \in \sigma(x)\}.$$

It follows that $\mathscr{H}(\sigma(x)) \ni f \mapsto \Phi(f)$ extends uniquely to an isometry $\mathsf{C}(\sigma(x)) \to \mathsf{B}(\mathcal{H})$ mapping any polynomial p to $p(x)$. But such a map must coincide with the continuous functional calculus and hence both calculi must agree on $\mathscr{H}(\sigma(x))$. □

Consider now the entire function exp. For any $x \in \mathsf{B}(\mathcal{H})$ the operator $\exp(x)$ can be written as the sum of the convergent series

$$\exp(x) = \sum_{m=0}^{\infty} \tfrac{1}{m!} x^m$$

(this follows from the definition of holomorphic functional calculus and expansion of exp in a power series). By standard manipulation performed on absolutely convergent series we easily check that if x and y commute then:

$$\exp(x)\exp(y) = \left(\sum_{n=0}^{\infty} \tfrac{x^n}{n!}\right)\left(\sum_{m=0}^{\infty} \tfrac{y^m}{m!}\right) = \sum_{k=0}^{\infty}\sum_{l=0}^{k} \tfrac{x^l}{l!}\tfrac{y^{k-l}}{(k-l)!}$$

$$= \sum_{k=0}^{\infty} \tfrac{1}{k!} \sum_{l=0}^{k} \binom{k}{l} x^l y^{k-l} = \sum_{k=0}^{\infty} \tfrac{(x+y)^k}{k!} = \exp(x+y).$$

The general situation is described by the next theorem:

Theorem 4.17 (Lie-Trotter Formula)
Let $x, y \in \mathsf{B}(\mathcal{H})$. Then

$$\exp(x+y) = \lim_{n\to\infty} \left(\exp\left(\tfrac{1}{n}x\right)\exp\left(\tfrac{1}{n}y\right)\right)^n.$$

Proof
For $n \in \mathbb{N}$ denote

$$s_n = \exp\left(\tfrac{1}{n}(x+y)\right), \qquad t_n = \exp\left(\tfrac{1}{n}x\right)\exp\left(\tfrac{1}{n}y\right).$$

We have

$$\|s_n\| = \left\| \sum_{m=0}^{\infty} \frac{1}{m!}\frac{1}{n^m}(x+y)^m \right\| \leq \sum_{m=0}^{\infty} \frac{1}{m!}\frac{1}{n^m}\left(\|x\| + \|y\|\right)^m = \exp\left(\frac{\|x\|+\|y\|}{n}\right)$$

and

$$\|t_n\| \leq \left\| \exp\left(\tfrac{1}{n}x\right)\right\| \left\| \exp\left(\tfrac{1}{n}y\right)\right\| \leq \exp\left(\frac{\|x\|}{n}\right)\exp\left(\frac{\|y\|}{n}\right) = \exp\left(\frac{\|x\|+\|y\|}{n}\right).$$

Furthermore

$$\begin{aligned}
\|s_n - t_n\| &= \left\| \sum_{m=0}^{\infty} \frac{1}{m!}\frac{1}{n^m}(x+y)^m - \left(\sum_{k=0}^{\infty} \frac{1}{k!}\frac{1}{n^k}x^k\right)\left(\sum_{l=0}^{\infty} \frac{1}{l!}\frac{1}{n^l}y^l\right)\right\| \\
&= \Bigg\| \mathbb{1} + \tfrac{1}{n}(x+y) + \sum_{m=2}^{\infty} \frac{1}{m!}\frac{1}{n^m}(x+y)^m \\
&\qquad - \mathbb{1} - \tfrac{1}{n}x - \tfrac{1}{n}y - \left(\sum_{k=1}^{\infty} \frac{1}{k!}\frac{1}{n^k}x^k\right)\left(\sum_{l=1}^{\infty} \frac{1}{l!}\frac{1}{n^l}y^l\right)\Bigg\| \\
&= \left\| \sum_{m=2}^{\infty} \frac{1}{m!}\frac{1}{n^m}(x+y)^m - \left(\sum_{k=1}^{\infty} \frac{1}{k!}\frac{1}{n^k}x^k\right)\left(\sum_{l=1}^{\infty} \frac{1}{l!}\frac{1}{n^l}y^l\right)\right\| \\
&= \frac{1}{n^2}\left\| \sum_{m=2}^{\infty} \frac{1}{m!}\frac{1}{n^{m-2}}(x+y)^m - \left(\sum_{k=1}^{\infty} \frac{1}{k!}\frac{1}{n^{k-1}}x^k\right)\left(\sum_{l=1}^{\infty} \frac{1}{l!}\frac{1}{n^{l-1}}y^l\right)\right\| \\
&= \frac{\text{const.}}{n^2}.
\end{aligned}$$

Using this and the identity

$$s_n^n - t_n^n = \sum_{r=0}^{n-1} s_n^r (s_n - t_n) t_n^{n-1-r}$$

we obtain the following estimate:

$$\begin{aligned}
\|s_n^n - t_n^n\| &\leq \sum_{r=0}^{n-1} \|s_n^r\| \|s_n - t_n\| \|t_n^{n-1-r}\| \\
&\leq n\Big(\max\left\{\|s_n\|, \|t_n\|\right\}\Big)^n \|s_n - t_n\| \leq \frac{\text{const.}}{n}\exp\left(\|x\| + \|y\|\right)
\end{aligned}$$

and hence

$$\lim_{n\to\infty}\left(\exp\left(\tfrac{1}{n}x\right)\exp\left(\tfrac{1}{n}y\right)\right)^n = \lim_{n\to\infty} t_n^n = \lim_{n\to\infty} s_n^n$$
$$= \lim_{n\to\infty}\left(\exp\left(\tfrac{1}{n}(x+y)\right)\right)^n = \exp(x+y).$$

□

4.5 Theorems of Fuglede and Putnam

Theorem 4.18 (Fuglede's Theorem)
Let x and y be commuting elements of $\mathsf{B}(\mathcal{H})$. Assume further that x is normal. Then $yx^* = x^*y$.

Proof
For $\lambda \in \mathbb{C}$ and let

$$u(\lambda) = \exp\left(\lambda x^* - \bar{\lambda}x\right).$$

Then $u(\lambda)$ is a unitary operator, since we can easily check that $u(\lambda)^* = u(-\lambda)$ for all λ and

$$u(\lambda)u(\mu) = u(\lambda+\mu), \qquad \lambda, \mu \in \mathbb{C}.$$

Moreover $u(0) = \mathbb{1}$ and

$$u(\lambda) = \exp(\lambda x^*)\exp\left(-\bar{\lambda}x\right) = \exp\left(-\bar{\lambda}x\right)\exp(\lambda x^*)$$

(as x is normal), so since x commutes with y, we see that

$$\exp(-\lambda x^*)y\exp(\lambda x^*) = \exp(-\lambda x^*)\exp\left(\bar{\lambda}x\right)y\exp\left(-\bar{\lambda}x\right)\exp(\lambda x^*)$$
$$= u(-\lambda)yu(\lambda).$$

Therefore the holomorphic function

$$h : \mathbb{C} \ni \lambda \longmapsto \exp(-\lambda x^*)y\exp(\lambda x^*) \in \mathsf{B}(\mathcal{H})$$

is bounded by $\|y\|$. By Liouville's theorem it is constant:

$$\exp(-\lambda x^*)y\exp(\lambda x^*) = h(0) = y, \qquad \lambda \in \mathbb{C}$$

which can be rewritten as

$$y\exp(\lambda x^*) = \exp(\lambda x^*)y,$$

i.e.

$$\sum_{n=0}^{\infty} \frac{\lambda^n}{n!} y(x^*)^n = \sum_{n=0}^{\infty} \frac{\lambda^n}{n!}(x^*)^n y.$$

Equality of convergent power series guarantees equality of all coefficients, so in particular $yx^* = x^*y$. □

Note that if $x \in \mathsf{B}(\mathcal{H})$ is not normal then the conclusion of Theorem 4.18 fails already for $y = x$.

Corollary 4.19 *Let $x_1, x_2, y \in \mathsf{B}(\mathcal{H})$. Assume that x_1 and x_2 are normal and that $yx_1 = x_2 y$. Then $yx_1{}^* = x_2{}^* y$.*

Proof
Identifying operators on $\mathcal{H} \oplus \mathcal{H}$ with 2×2 matrices over $\mathsf{B}(\mathcal{H})$ consider operators $\widetilde{x}$ and $\widetilde{y}$ on $\mathcal{H} \oplus \mathcal{H}$ given by

$$\widetilde{x} = \begin{bmatrix} x_1 & 0 \\ 0 & x_2 \end{bmatrix} \quad \text{and} \quad \widetilde{y} = \begin{bmatrix} 0 & 0 \\ y & 0 \end{bmatrix}.$$

Then $\widetilde{x}$ is normal and we have $\widetilde{y}\widetilde{x} = \widetilde{x}\widetilde{y}$. By Fuglede's theorem $\widetilde{y}\widetilde{x}^* = \widetilde{x}^*\widetilde{y}$ which means that $yx_1{}^* = x_2{}^* y$. □

We say that two operators $x_1, x_2 \in \mathsf{B}(\mathcal{H})$ are *similar* if there exists an invertible $y \in \mathsf{B}(\mathcal{H})$ such that $yx_1 y^{-1} = x_2$. Corollary 4.19 allows us to prove that similar normal operators are, in fact, unitarily equivalent.

Corollary 4.20 (Putnam's Theorem) *Let $x_1, x_2 \in \mathsf{B}(\mathcal{H})$ be normal. If x_1 and x_2 are similar then they are unitarily equivalent.*

Proof
There exists an invertible $y \in \mathsf{B}(\mathcal{H})$ such that $yx_1 y^{-1} = x_2$, i.e. $yx_1 = x_2 y$. By the previous corollary we have $yx_1{}^* = x_2{}^* y$ which can be rewritten as

$$x_1 y^* = y^* x_2.$$

From this we infer that

$$y^* y x_1 = y^* x_2 y = x_1 y^* y,$$

and therefore also $|y| = (y^*y)^{\frac{1}{2}}$ commutes with x_1 (cf. Remark 2.7). Let $y = u|y|$ be the polar decomposition of y. Since both y and $|y|$ are invertible, u is an invertible partial isometry. Hence u is unitary and, moreover, we have

$$ux_1u^* = u|y||y|^{-1}x_1|y||y|^{-1}u^{-1} = yx_1y^{-1} = x_2.$$

□

Putnam's theorem can be slightly strengthened in the following way: assume $x_1, x_2 \in \mathsf{B}(\mathcal{H})$ are normal and $y \in \mathsf{B}(\mathcal{H})$ is such that $yx_1 = x_2y$. Then, precisely as above we show that $|y|x_1 = x_1|y|$. Thus

$$x_2u|y| = x_2y = yx_1 = u|x|x_1 = ux_1|y|. \tag{4.7}$$

Now if $\ker y = \{0\}$ then also $\ker|y| = \{0\}$ and $\overline{|y|\mathcal{H}} = (\ker y)^\perp = \mathcal{H}$, so it follows from (4.7) that $x_2u = ux_1$. If we additionally assume that the range of y is dense in $\mathcal{H}$ then u is unitary (Remark 3.14(1)) and so it implements unitary equivalence of x_1 and x_2. In other words we have:

Corollary 4.21 *Let $x_1, x_2 \in \mathsf{B}(\mathcal{H})$ be normal. If there exists $y \in \mathsf{B}(\mathcal{H})$ such that $yx_1 = x_2y$, $\ker y = \{0\}$ and $\overline{y\mathcal{H}} = \mathcal{H}$ then x_1 and x_2 are unitarily equivalent.*

4.6 Functional Calculus in C*-algebras

Let A be a C*-algebra with unit. An element $a \in \mathsf{A}$ is called *self-adjoint* if $a = a^*$, while we say that a is *normal* if $aa^* = a^*a$. Continuing the analogy with $\mathsf{B}(\mathcal{H})$ we say that $u \in \mathsf{A}$ is *unitary* if $u^*u = uu^* = \mathbb{1}$. Notice that if $u \in \mathsf{A}$ is unitary and $\lambda \in \sigma(u)$ (spectrum of a C*-algebra element was defined in ▶ Sect. 1.3) then $|\lambda| = 1$. Indeed: first of all we have $\lambda^{-1} \in \sigma(u^{-1})$, since if $\lambda^{-1}\mathbb{1} - u^{-1}$ were invertible then so would be $\lambda\mathbb{1} - u$, as

$$(\lambda\mathbb{1} - u)\big(-\lambda^{-1}u^{-1}(\lambda^{-1}\mathbb{1} - u^{-1})^{-1}\big)$$
$$= \big(-\lambda^{-1}u^{-1}(\lambda^{-1}\mathbb{1} - u^{-1})^{-1}\big)(\lambda\mathbb{1} - u) = \mathbb{1}.$$

Furthermore, since $\lambda^{-1} \in \sigma(u^*) = \overline{\sigma(u)} \subset \big\{z \in \mathbb{C}\,\big|\,|z| \le 1\big\}$ and $\lambda \in \big\{z \in \mathbb{C}\,\big|\,|z| \le 1\big\}$, we find that $|\lambda| = 1$.

For any element $a \in \mathsf{A}$ and a function f holomorphic on a neighborhood of $\sigma(a)$ we put

$$f(a) = \tfrac{1}{2\pi i}\oint_\Gamma f(\lambda)(\lambda\mathbb{1} - a)^{-1}d\lambda,$$

where Γ is a positively oriented curve surrounding $\sigma(a)$ contained within the domain of holomorphy of f. This way we define a unital homomorphism from the algebra of functions holomorphic on a neighborhood of $\sigma(a)$ into A which maps polynomials onto corresponding polynomials in a. Moreover, for a and f as above we have $\sigma\big(f(a)\big) = f\big(\sigma(a)\big)$. Proofs of these facts are identical to those for $\mathsf{A} = \mathsf{B}(\mathcal{H})$.

Proposition 4.22 *Let a be an element of a unital* C*-*algebra* A. *Then*
(1) *if a is normal then* $|\sigma(a)| = \|a\|$,
(2) *if a is self-adjoint then* $\sigma(a) \subset \mathbb{R}$.

Proof
The proof of (1) is established in exactly the same way as that of Proposition 2.3(1). Now let a be self-adjoint. Then we easily check that the element $u = \exp(ia)$ is unitary. By holomorphic functional calculus for a we see that $\lambda \in \sigma(a)$ implies $\exp(i\lambda) \in \sigma(u)$, so that $\big|\exp(i\lambda)\big| = 1$, and hence $\lambda \in \mathbb{R}$. □

Similarly as for holomorphic functions, we can also generalize the continuous functional calculus for operators to self-adjoint elements of C*-algebras:

Theorem 4.23
Let $a \in \mathsf{A}$ *be self-adjoint. Then there exists a unique map*

$$\mathsf{C}\big(\sigma(a)\big) \ni f \longmapsto f(a) \in \mathsf{A}$$

such that
- *if f is a polynomial function* $f(\lambda) = \alpha_0 + \alpha_1\lambda + \cdots + \alpha_n\lambda^n$ *then*

$$f(a) = \alpha_0\mathbb{1} + \alpha_1 a + \cdots + \alpha_n a^n,$$

- $\big\|f(a)\big\| = \|f\|_\infty$ *for all* $f \in \mathsf{C}\big(\sigma(a)\big)$.

Moreover, this map is a $*$*-isomorphism of the* C*-*algebra* $\mathsf{C}\big(\sigma(a)\big)$ *onto the smallest closed* $*$*-subalgebra of* A *containing* $\mathbb{1}$ *and a.*

The proof of Theorem 4.23 is fully analogous to that of existence and uniqueness of continuous functional calculus for self-adjoint operators. We also have the following:

Proposition 4.24 *Let* A *be a unital* C*-*algebra and let* $a \in \mathsf{A}$ *be self-adjoint. Then*
(1) *if b is a self-adjoint element of* A *commuting with a then for any functions* $f \in \mathsf{C}\big(\sigma(a)\big)$ *and* $g \in \mathsf{C}\big(\sigma(b)\big)$ *we have* $f(a)g(b) = g(b)f(a)$,
(2) *if* $c \in \mathsf{A}$ *commutes with a then for any* $f \in \mathsf{C}\big(\sigma(a)\big)$ *we have* $f(a)c = cf(a)$,
(3) *for any* $f \in \mathsf{C}\big(\sigma(a)\big)$ *we have* $\sigma\big(f(a)\big) = f\big(\sigma(a)\big)$,
(4) *for any real-valued* $g \in \mathsf{C}\big(\sigma(a)\big)$ *the element* $g(a) \in \mathsf{A}$ *is self-adjoint and for any* $f \in \mathsf{C}\big(g\big(\sigma(a)\big)\big)$ *we have* $f\big(g(a)\big) = (f \circ g)(a)$.

As before, proofs of the facts listed above amount to repeating the arguments used for self-adjoint operators on a Hilbert space.

An element $a \in \mathsf{A}$ is called *positive* if a is self-adjoint and $\sigma(a)$ is contained in $[0, +\infty[$. Using functional calculus we can easily show that any positive element has a unique positive square root. The notion of a positive element introduces a partial order on A: by definition $a \leq b$ if and only if $b - a$ is positive.

Proposition 4.25 *Let $a, b \in \mathsf{A}$ be elements such that $0 \leq a \leq b$. Assume that a is invertible. Then b is invertible and $b^{-1} \leq a^{-1}$.*

Once more the proof of this proposition is identical to that given for operators (Proposition 3.7).

Notes

Further information on multiplication operators and Borel functional calculus can be found in the textbooks [Arv_2, ReSi_1, Rud_2] and in [Hal, Section 7]. A more general view of holomorphic functional calculus from ▶ Sect. 4.6 is presented in the monograph [Zel] and in [Arv_1]. There, the reader will also find various examples and problems related to this topic.

Theorem 4.7 as well as all results of ▶ Sects. 4.2 and 4.3 for self-adjoint operators can be extended to apply to all normal operators. As we already mentioned in ▶ Chap. 2, these extensions will be presented in ▶ Sect. 7.4. Furthermore, functional calculus in C*-algebras also extends to all normal elements, so that Theorem 4.23 and Proposition 4.24 hold for normal elements without substantial changes. However, the proofs of these results are best accomplished with techniques different from the ones used above for self-adjoint elements.

Compact Operators

P. Sołtan, *A Primer on Hilbert Space Operators*, Compact Textbooks in Mathematics,
https://doi.org/10.1007/978-3-319-92061-0_5

In this chapter we will discuss the most fundamental results on compact operators on Hilbert spaces. It is important to stress that analogous theory for operators on Banach spaces is somewhat more complicated and, in particular, requires more considerations of topological nature. The very definition of a compact operator between Banach spaces is significantly different from the one we will propose in the case of operators on a Hilbert space. In the more general framework of Banach spaces the definition of a compact operator amounts to the conclusion of Proposition 5.1. An interested reader will find more information on compact operators in any functional analysis textbook, e.g. in [Rud_2, Chapter 4].

5.1 Compact Operators on a Hilbert Space

We say that an operator $x \in \mathsf{B}(\mathcal{H})$ is *finite dimensional* if its range is finite a dimensional subspace of $\mathcal{H}$. A *compact* operator on $\mathcal{H}$ is an element $x \in \mathsf{B}(\mathcal{H})$ which is a norm limit of finite dimensional operators. The sets of finite dimensional and compact operators will be denoted by $\mathcal{F}(\mathcal{H})$ and $\mathsf{B}_0(\mathcal{H})$ respectively. It is easy to check that both $\mathcal{F}(\mathcal{H})$ and $\mathsf{B}_0(\mathcal{H})$ are *ideals* in $\mathsf{B}(\mathcal{H})$, i.e. they are vector subspaces and the implications

$$\Big(x \in \mathcal{F}(\mathcal{H}),\ y \in \mathsf{B}(\mathcal{H}) \Big) \Longrightarrow \Big(xy, yx \in \mathcal{F}(\mathcal{H}) \Big)$$

and

$$\Big(x \in \mathsf{B}_0(\mathcal{H}),\ y \in \mathsf{B}(\mathcal{H}) \Big) \Longrightarrow \Big(xy, yx \in \mathsf{B}_0(\mathcal{H}) \Big)$$

hold.

Any finite dimensional operator x is of the form

$$x\xi = \sum_{n=1}^{N} \alpha_n(\xi)\psi_n, \qquad \xi \in \mathcal{H}$$

for some $\psi_1, \ldots, \psi_N \in \mathcal{H}$ (basis of the range of x) and continuous linear functionals $\alpha_1, \ldots, \alpha_N$ on $\mathcal{H}$. It follows that $\alpha_i(\xi) = \langle \varphi_i | \xi \rangle$ for some $\varphi_1, \ldots, \varphi_n \in \mathcal{H}$, so that

$$x = \sum_{n=1}^{N} |\psi_n\rangle \langle \varphi_n| .$$

Thus

$$x^* = \sum_{n=1}^{N} |\varphi_n\rangle \langle \psi_n|$$

is also finite dimensional, and hence the ideal $\mathcal{F}(\mathcal{H})$ is closed under the involution of $\mathsf{B}(\mathcal{H})$. Consequently the ideal $\mathsf{B}_0(\mathcal{H})$ is closed under taking adjoints, since if $x = \lim_{n\to\infty} x_n$ and $x_n \in \mathcal{F}(\mathcal{H})$ for all n then $x^* = \lim_{n\to\infty} x_n{}^* \in \mathsf{B}_0(\mathcal{H})$. The ideal $\mathsf{B}_0(\mathcal{H})$ is moreover norm-closed. In particular if $x \in \mathsf{B}_0(\mathcal{H})$ is self-adjoint and f is a continuous function on $\sigma(x)$ which can be uniformly approximated by polynomials without constant term then $f(x) \in \mathsf{B}_0(\mathcal{H})$.

Proposition 5.1 *An operator $x \in \mathsf{B}(\mathcal{H})$ is compact if and only if for any bounded set $S \subset \mathcal{H}$ the set $x(S)$ is pre-compact.*[1]

Proof

Let us denote the closed unit ball of $\mathcal{H}$ by the symbol $\mathcal{H}_1$. Clearly it is enough to prove that x is compact if and only if $x(\mathcal{H}_1)$ is a pre-compact set.

Let x be a compact operator and fix $\varepsilon > 0$. There exists a sequence $(x_n)_{n\in\mathbb{N}}$ of finite dimensional operators such that $x_n \xrightarrow[n\to\infty]{} x$, and so there exists a finite dimensional operator x_n such that $\|x - x_n\| < \frac{\varepsilon}{2}$. The set $x_n(\mathcal{H}_1)$ is pre-compact (it is bounded and contained in a finite dimensional subspace), so it has a finite $\frac{\varepsilon}{2}$-net $\{\eta_1, \ldots, \eta_N\}$. Now for $\xi \in \mathcal{H}_1$ we have

$$\|x\xi - \eta_k\| \leq \|x\xi - x_n\xi\| + \|x_n\xi - \eta_k\|, \qquad k \in \{1, \ldots, N\}.$$

The term $\|x\xi - x_n\xi\|$ is bounded by $\frac{\varepsilon}{2}$, and since $x_n\xi$ belongs to $x_n(\mathcal{H}_1)$, there exists $k_0 \in \{1, \ldots, N\}$ such that $\|x_n\xi - \eta_{k_0}\| < \frac{\varepsilon}{2}$. Thus $\|x\xi - \eta_{k_0}\| < \varepsilon$, and so $\{\eta_1, \ldots, \eta_N\}$ is an ε-net for $x(\mathcal{H}_1)$.

[1] A subset of a topological space is *pre-compact* if its closure is compact. A subset S of a complete metric space is pre-compact if and only if for any $\varepsilon > 0$ there exists a finite family of balls with radius ε covering S. The collection of centers of such balls is called an *ε-net* for S.

Now suppose that $x \in \mathrm{B}(\mathcal{H})$ is such that for any $\varepsilon > 0$ there is a finite ε-net $\{\eta_1^\varepsilon, \ldots, \eta_N^\varepsilon\}$ for $x(\mathcal{H}_1)$. Let p_ε be the projection onto the finite dimensional subspace

$$\operatorname{span}\{\eta_1^\varepsilon, \ldots, \eta_N^\varepsilon\}$$

and let $x_\varepsilon = p_\varepsilon x$. Then $x_\varepsilon \in \mathcal{F}(\mathcal{H})$ and for any $\xi \in \mathcal{H}_1$ there exists k such that

$$\|x\xi - \eta_k^\varepsilon\| < \varepsilon.$$

Therefore

$$\begin{aligned}\|x\xi - x_\varepsilon\xi\| = \big\|(\mathbb{1} - p_\varepsilon)x\xi\big\| &\leq \big\|(\mathbb{1} - p_\varepsilon)(x\xi - \eta_k^\varepsilon)\big\| + \big\|(\mathbb{1} - p_\varepsilon)\eta_k^\varepsilon\big\| \\ &= \big\|(\mathbb{1} - p_\varepsilon)(x\xi - \eta_k^\varepsilon)\big\| \leq \|\mathbb{1} - p_\varepsilon\|\|x\xi - \eta_k^\varepsilon\| < \varepsilon,\end{aligned}$$

as the vector η_k^ε is orthogonal to the range of the projection $\mathbb{1} - p_\varepsilon$.

This way we have shown that for any ε there exists a finite dimensional operator within distance less than ε of x. It follows easily from this that x is a norm limit of finite dimensional operators. □

Let $x \in \mathrm{B}(\mathcal{H})$ be compact and suppose that $\lambda \in \sigma(x)$ is a non-zero eigenvalue. Then the eigenspace $\mathcal{K} = \big\{\xi \in \mathcal{H} \bigm| x\xi = \lambda\xi\big\}$ for λ must be finite dimensional. Indeed: the image under x of the bounded set $\mathcal{K}_1 = \big\{\xi \in \mathcal{K} \bigm| \|\xi\| \leq 1\big\}$ is the ball $\big\{\xi \in \mathcal{K} \bigm| \|\xi\| \leq |\lambda|\big\}$ which is not pre-compact unless $\dim \mathcal{K} < +\infty$. This shows, in particular, that the operator $\mathbb{1}$ on an infinite dimensional Hilbert space is not compact. More generally a projection is compact if and only if it is finite dimensional. On the other hand the next proposition says that there are in a sense many compact operators.

Proposition 5.2 *For any Hilbert space $\mathcal{H}$ there exists a net of finite dimensional projections converging to $\mathbb{1}$ in strong topology.*

Proof

Let $\{\psi_i\}_{i\in I}$ be an orthonormal basis of $\mathcal{H}$ and let J be the set of finite subsets of I directed by inclusion. For $j \in J$ let

$$p_j = \sum_{i\in j} |\psi_i\rangle\langle\psi_i|.$$

Then for any $\xi \in \mathcal{H}$ we have $\xi = \sum_{i\in I} \langle\psi_i|\xi\rangle\,\psi_i$ and since the coefficients of ξ in the basis $\{\psi_i\}_{i\in I}$ are square-summable, we have

$$\|p_j\xi - \xi\|^2 = \sum_{i\in I\setminus j} \big|\langle\psi_i|\xi\rangle\big|^2 \xrightarrow[j\in J]{} 0.$$

□

5.2 Fredholm Alternative

Theorem 5.3
Let D be a connected open subset of $\mathbb{C}$ and let $f : D \to \mathsf{B}(\mathcal{H})$ be a holomorphic function all of whose values are compact operators. Then either
(a) *for any $z \in D$ the operator $\mathbb{1} - f(z)$ is not invertible*
or
(b) *the operator $\mathbb{1} - f(z)$ is invertible for z belonging to $D \setminus S$, where S is a subset of D without an accumulation point in D; in this case for any $z \in S$ the equation $f(z)\psi = \psi$ has a non-zero solution.*

Proof
We note first that it is enough to prove that either (a) or (b) holds on a neighborhood of any given $z_0 \in D$.

Fix $z_0 \in D$ and let $r > 0$ be such that

$$D_r = \{z \in \mathbb{C} \,|\, |z - z_0| < r\}$$

is contained in D and $z \in D_r$ implies that $\|f(z) - f(z_0)\| < \frac{1}{2}$. Let y be a finite dimensional operator such that $\|f(z_0) - y\| < \frac{1}{2}$. Then for any $z \in D_r$ we have $\|f(z) - y\| < 1$, and so $\mathbb{1} - f(z) + y$ is invertible and the function

$$D_r \ni z \longmapsto \left(\mathbb{1} - f(z) + y\right)^{-1}$$

is holomorphic.

The operator y can be written as

$$y = \sum_{n=1}^{N} |\psi_n\rangle\langle\varphi_n| \tag{5.1}$$

for some vectors $\varphi_1, \ldots, \varphi_N, \psi_1, \ldots \psi_N \in \mathcal{H}$ and we can assume that the system $\{\psi_1, \ldots, \psi_N\}$ is linearly independent. For $z \in D_r$ let

$$\varphi_n(z) = \left(\left(\mathbb{1} - f(z) + y\right)^{-1}\right)^* \varphi_n$$

and

$$g(z) = y\left(\mathbb{1} - f(z) + y\right)^{-1}.$$

Using (5.1) we find that

$$\begin{aligned} g(z) &= \sum_{n=1}^{N} |\psi_n\rangle \langle\varphi_n| \big(\mathbb{1} - f(z) + y\big)^{-1} \\ &= \sum_{n=1}^{N} \Big|\psi_n\Big\rangle\Big\langle\big((\mathbb{1} - f(z) + y)^{-1}\big)^{*}\varphi_n\Big| \\ &= \sum_{n=1}^{N} |\psi_n\rangle \langle\varphi_n(z)| . \end{aligned} \tag{5.2}$$

It follows that for any $z \in D_r$ we have $g(z)\mathcal{H} \subset y\mathcal{H}$.

It is clear that

$$\big(\mathbb{1} - g(z)\big)\big(\mathbb{1} - f(z) + y\big) = \mathbb{1} - f(z) + y - y = \mathbb{1} - f(z),$$

and since the operator $\mathbb{1} - f(z) + y$ is invertible for $z \in D_r$, we see that

- $\mathbb{1} - f(z)$ is invertible if and only if $\mathbb{1} - g(z)$ is invertible,
- the equation $\psi = f(z)\psi$ has a non-zero solution (i.e. $\ker\big(\mathbb{1} - f(z)\big) \neq \{0\}$) if and only if the equation $\varphi = g(z)\varphi$ has a non-zero solution ($\ker\big(\mathbb{1} - g(z)\big) \neq \{0\}$).

Suppose that $\varphi = g(z)\varphi$. Then φ belongs to the range of $g(z)$ which is contained in the range of y, so that $\varphi = \sum_{m=1}^{N} \beta_m \psi_m$. In view of (5.2) we get

$$\beta_n = \sum_{m=1}^{N} \langle\varphi_n(z)|\psi_m\rangle\, \beta_m . \tag{5.3}$$

Conversely, if $\beta_1, \ldots, \beta_N$ satisfy (5.3) then $\varphi = \sum_{m=1}^{N} \beta_m \psi_m$ is a solution to the equation $\varphi = g(z)\varphi$. It follows that this equation has a non-zero solution if and only if

$$d(z) = \det\left(\begin{bmatrix} 1 & & \\ & \ddots & \\ & & 1 \end{bmatrix} - \begin{bmatrix} \langle\varphi_1(z)|\psi_1\rangle & \ldots & \langle\varphi_1(z)|\psi_N\rangle \\ \vdots & \ddots & \vdots \\ \langle\varphi_N(z)|\psi_1\rangle & \ldots & \langle\varphi_N(z)|\psi_N\rangle \end{bmatrix}\right) = 0.$$

The function d is holomorphic on D_r, and consequently the set $S_r = \big\{z \in D_r \,\big|\, d(z) = 0\big\}$ either has no accumulation points in D_r or $S_r = D_r$.

Let us now investigate invertibility of $\mathbb{1} - g(z)$. For this we note that the equation

$$\big(\mathbb{1} - g(z)\big)\varphi = \xi \tag{5.4}$$

has a solution for any $\xi \in \mathcal{H}$ if and only if the equation $\varphi - g(z)\varphi = \zeta$ has a solution for any ζ in the range of $g(z)$. Indeed: substituting

$$\varphi = \xi + \varphi' \tag{5.5}$$

in (5.4) we obtain

$$\xi + \varphi' - g(z)\xi - g(z)\varphi' = \xi,$$

hence

$$\varphi' - g(z)\varphi' = g(z)\xi. \tag{5.6}$$

If φ' is a solution to this equation then $\varphi = \xi + \varphi'$ is a solution of (5.4), and conversely, if φ satisfies (5.4) then $\varphi' = \varphi - \xi$ satisfies (5.6). Moreover, as ξ runs over all of $\mathcal{H}$, the vector $\zeta = g(z)\xi$ runs over the range of $g(z)$. In other words, we can solve (5.4) for any ξ if and only if $d(z) \neq 0$. Summing up, for $z \in D_r$

$$d(z) = 0 \Longleftrightarrow \Big(\text{ there exists } \varphi \neq 0 \text{ such that } \varphi = g(z)\varphi \Big)$$

and

$$d(z) \neq 0 \Longleftrightarrow \Big(\text{ for any } \xi \in \mathcal{H} \text{ there exists } \varphi \text{ such that } \big(\mathbb{1} - g(z)\big)\varphi = \xi \Big).$$

Thus $\mathbb{1} - g(z)$ is either not invertible on all of D_r (when $S = D_r$) or S_r is a discrete subset of D_r and $\mathbb{1} - g(z)$ is a bijection $\mathcal{H} \to \mathcal{H}$ for $z \in D_r \setminus S_r$. In the latter case, for $z \in D_r \setminus S_r$ we have a formula for $\big(\mathbb{1} - g(z)\big)^{-1}$, namely

$$\big(\mathbb{1} - g(z)\big)^{-1}\xi = \Big(\mathbb{1} + \big(\big(\mathbb{1} - g(z)\big)\big|_{y\mathcal{H}}\big)^{-1} g(z)\Big)\xi$$

(cf. (5.5)) which shows that $\big(\mathbb{1} - g(z)\big)^{-1}$ is bounded. □

Corollary 5.4 (Fredholm Alternative) *Let $x \in \mathrm{B}_0(\mathcal{H})$. Then either $\mathbb{1} - x$ is invertible or the equation $\psi = x\psi$ has a non-zero solution.*

Proof
Put $D = \mathbb{C}$ and $f(z) = zx$. Then the statement follows from Theorem 5.3 for $z = 1$. □

Corollary 5.5 (Riesz-Schauder Theorem) *Let $x \in \mathrm{B}_0(\mathcal{H})$. Then the set $\sigma(x)$ does not have accumulation points other than possibly $\lambda = 0$. Moreover any non-zero element of $\sigma(x)$ is an eigenvalue of finite multiplicity.*

Proof
As in the proof of Corollary 5.4 we apply Theorem 5.3 to the function $f(x) = zx$ on $\mathbb{C}$. It follows that the set

$$S = \{z \in \mathbb{C} \,|\, \text{the equation } zx\psi = \psi \text{ has a non-zero solution}\}$$

is discrete (it is not equal to $\mathbb{C}$, as $0 \notin S$). Moreover, for z outside of S the operator $\mathbb{1} - zx$ is invertible, i.e. the operator

$$\lambda\mathbb{1} - x = \lambda(\mathbb{1} - \lambda^{-1}x)$$

is invertible provided $\lambda \neq 0$ and λ lies outside the discrete set

$$S^{-1} = \{z^{-1} \,|\, z \in S\} \subset \mathbb{C}\setminus\{0\},$$

and any $\lambda \in S^{-1}$ is an eigenvalue of x. The fact that non-zero eigenvalues of a compact operator must have finite multiplicity has already been discussed at the end of ▶ Sect. 5.1. □

Note that since a discrete subset of $\mathbb{C}$ is necessarily countable, spectrum of a compact operator is a countable set.

Corollary 5.6 (Hilbert-Schmidt Theorem) *Let $x \in \mathrm{B}(\mathcal{H})$ be a compact normal operator. Then there exists an orthonormal basis of $\mathcal{H}$ consisting of eigenvectors of x.*

Proof
Write the spectrum of x as

$$\sigma(x) = \{0\} \cup \{\lambda_1, \lambda_2, \ldots\},$$

where for any $n \in \mathbb{N}$ we have $\lambda_n \neq 0$. Let $\{\psi_j^0\}_{j\in J}$ be an orthonormal basis of $\ker x$ and for $n \in \mathbb{N}$ let $\{\psi_k^n\}_{k=1,\ldots,\dim\ker(\lambda_n\mathbb{1}-x)}$ be an orthonormal basis of the (finite dimensional) eigenspace of x for the eigenvalue λ_n. The union $\{\varphi_i\}_{i\in I}$ of all these bases is an orthonormal system in $\mathcal{H}$. Define the subspace $\mathcal{S} = \overline{\mathrm{span}}\{\varphi_i \,|\, i \in I\}$.

Both x and x^* leave $\mathcal{S}$ invariant, and hence they also leave $\mathcal{S}^\perp$ invariant (cf. Proposition 2.1). Furthermore the operator $x\big|_{\mathcal{S}^\perp}$ is compact, and consequently non-zero elements of its spectrum must be eigenvalues. But all eigenvectors of x by definition belong to $\mathcal{S}$, so $\sigma\left(x\big|_{\mathcal{S}^\perp}\right) = \{0\}$. As $x\big|_{\mathcal{S}^\perp}$ is also normal, we have $x\big|_{\mathcal{S}^\perp} = 0$ because its spectral radius is equal to 0. This, however, shows that $\mathcal{S}^\perp$ consists solely of eigenvectors of x (for the eigenvalue 0), so $\mathcal{S}^\perp \subset \mathcal{S}$. It follows that $\mathcal{S}^\perp = \{0\}$ or, in other words, $\mathcal{S} = \mathcal{H}$ and we find that $\{\varphi_i\}_{i\in I}$ is an orthonormal basis of $\mathcal{H}$. □

Suppose a compact operator x has infinite spectrum and let $\{\lambda_n\}_{n\in\mathbb{N}}$ be the sequence of all non-zero eigenvalues of x. Then $\lambda_n \xrightarrow[n\to\infty]{} 0$. Indeed: for any $r > 0$ the set $\{n \mid |\lambda_n| > r\}$ is finite because for any n we have $|\lambda_n| \leq \|x\|$, so that $\{\lambda_n \mid |\lambda_n| \geq r\}$ is a sequence of elements of a compact set $\{z \in \mathbb{C} \mid r \leq |z| \leq \|x\|\}$ which cannot be infinite, since it does not have accumulation points.

Corollary 5.7 (Canonical Form of a Compact Operator) *Let $x \in \mathsf{B}(\mathcal{H})$ be a compact operator. Then there exists a finite or countably infinite set N and orthonormal systems $\{\psi_n\}_{n\in N}$ and $\{\varphi_n\}_{n\in N}$ in $\mathcal{H}$ as well as a sequence $\{\lambda_n\}_{n\in N}$ of strictly positive numbers such that*

$$x = \sum_{n\in N} \lambda_n |\varphi_n\rangle \langle\psi_n|,$$

with the sum convergent in norm.

Proof

Let $x = v|x|$ be the polar decomposition of x. Then the operator $|x| = (x^*x)^{\frac{1}{2}}$ is compact and positive. Therefore there exists an orthonormal basis of $\mathcal{H}$ consisting of eigenvectors of $|x|$. Let $\{\psi_n\}_{n\in N}$ be the subsystem of this basis consisting of eigenvectors corresponding to non-zero eigenvalues. Then $\{\psi_n\}_{n\in N}$ is an orthonormal basis of $\left(\ker|x|\right)^{\perp}$. We have

$$|x| = \sum_{n\in N} \lambda_n |\psi_n\rangle \langle\psi_n|$$

and the sum is norm convergent because for a finite $F \subset N$ we have

$$\sum_{n\in F} \lambda_n |\psi_n\rangle \langle\psi_n| = f_F\left(|x|\right),$$

where f_F is the continuous function on $\sigma\left(|x|\right) = \overline{\{\lambda_n\}_{n\in N}}$ given by

$$f_F(\lambda) = \begin{cases} \lambda & \lambda \in F, \\ 0 & \lambda \notin F \end{cases}$$

and f_F converges uniformly to the identity function on $\sigma\left(|x|\right)$ as F ranges over the directed set of finite subsets of N (cf. remarks preceding Corollary 5.7).

The partial isometry v maps the subspace $\left(\ker|x|\right)^{\perp}$ isometrically onto $\overline{x\mathcal{H}}$. Therefore putting $\varphi_n = v\psi_n$ for all $n \in N$ we obtain another orthonormal system $\{\varphi_n\}_{n\in N}$. Clearly

$$x = v\sum_{n\in N} \lambda_n |\psi_n\rangle \langle\psi_n| = \sum_{n\in N} \lambda_n |\varphi_n\rangle \langle\psi_n|.$$

□

Non-zero eigenvalues of $|x|$ appearing in the canonical form of a compact operator x are called the *singular values* of the operator x. Note that the orthonormal systems in the canonical form of x are far from being unique.

Notes

The study of compact operators between Banach spaces constituted the foundation of functional analysis and its applications with first examples of such operators coming from the theory of integral and differential equations. Examples of applications of the theory of compact operators in these areas of mathematics can be found e.g. in [Mau, Chapter VII], [Ped, Section 3.3], [ReSi$_1$, Chapter VI], while interesting problems and other examples concerning compact operators are available in the problem book [Hal].

The Trace

P. Sołtan, *A Primer on Hilbert Space Operators*, Compact Textbooks in Mathematics,
https://doi.org/10.1007/978-3-319-92061-0_6

We will now discuss the concept of the trace of an operator. This notion has applications in quantum physics as well as in many problems of pure mathematics. One of them concerns a precise description of the dual space of $\mathsf{B}_0(\mathcal{H})$ which we will give in detail. Analysis of the trace as an "unbounded linear functional" on $\mathsf{B}(\mathcal{H})$ leads to far reaching generalizations in the theory of operator algebras (under the guise of theory of weights). Our presentation of this topic will be based on the same foundations which underlie those generalizations and can be treated as an introduction to more advanced techniques of the theory of operator algebras.

6.1 Definition of the Trace

Let $\mathcal{H}$ be a Hilbert space and let $\{\xi_j\}_{j\in J}$ be an orthonormal basis of $\mathcal{H}$. For a positive $t \in \mathsf{B}(\mathcal{H})$ define

$$\mathrm{Tr}(t) = \sum_{j\in J} \langle \xi_j | t\xi_j \rangle \in [0, +\infty].$$

Proposition 6.1 *For any* $x \in \mathsf{B}(\mathcal{H})$ *we have* $\mathrm{Tr}(x^*x) = \mathrm{Tr}(xx^*)$.

Proof
For each $i \in J$ we have

$$\begin{aligned}\sum_{j\in J} \langle \xi_i | x^*\xi_j \rangle \langle \xi_j | x\xi_i \rangle &= \sum_{j\in J} \langle x\xi_i | \xi_j \rangle \langle \xi_j | x\xi_i \rangle \\ &= \left\langle x\xi_i \,\middle|\, \sum_{j\in J} \langle \xi_j | x\xi_i \rangle \xi_j \right\rangle \\ &= \langle x\xi_i | x\xi_i \rangle = \langle \xi_i | x^*x\xi_i \rangle,\end{aligned} \tag{6.1}$$

and hence

$$\mathrm{Tr}(x^*x) = \sum_{i\in J} \langle \xi_i | x^*x\xi_i \rangle = \sum_{i\in J}\sum_{j\in J} \langle \xi_i | x^*\xi_j \rangle \langle \xi_j | x\xi_i \rangle .$$

Now, since

$$\langle \xi_i | x^*\xi_j \rangle \langle \xi_j | x\xi_i \rangle = \langle x\xi_i | \xi_j \rangle \langle \xi_j | x\xi_i \rangle = \left| \langle \xi_j | x\xi_i \rangle \right|^2 \geq 0,$$

we can change the order of summation and using the calculation (6.1) in the fourth step (this time applying it to the sum over i and the operator x^*), we obtain

$$\begin{aligned}
\mathrm{Tr}(x^*x) &= \sum_{i\in J}\sum_{j\in J} \langle \xi_i | x^*\xi_j \rangle \langle \xi_j | x\xi_i \rangle \\
&= \sum_{j\in J}\sum_{i\in J} \langle \xi_i | x^*\xi_j \rangle \langle \xi_j | x\xi_i \rangle \\
&= \sum_{j\in J}\sum_{i\in J} \langle \xi_j | x\xi_i \rangle \langle \xi_i | x^*\xi_j \rangle \\
&= \sum_{j\in J} \langle \xi_j | xx^*\xi_j \rangle = \mathrm{Tr}(xx^*).
\end{aligned}$$

□

Corollary 6.2 *Let $u, t \in \mathsf{B}(\mathcal{H})$ with u unitary and t positive. Then*

$$\mathrm{Tr}(utu^*) = \mathrm{Tr}(t).$$

Proof

Put $x = ut^{\frac{1}{2}}$. Then

$$x^*x = t^{\frac{1}{2}}u^*ut^{\frac{1}{2}} = t \quad \text{and} \quad xx^* = ut^{\frac{1}{2}}t^{\frac{1}{2}}u^* = utu^*.$$

By Proposition 6.1 we have $\mathrm{Tr}(utu^*) = \mathrm{Tr}(xx^*) = \mathrm{Tr}(x^*x) = \mathrm{Tr}(t)$. □

Corollary 6.3 *Let $\{\eta_j\}_{j\in J}$ be another orthonormal basis of $\mathcal{H}$ and define*

$$\mathrm{Tr}'(t) = \sum_{j\in J} \langle \eta_j | t\eta_j \rangle, \qquad t \in \mathsf{B}(\mathcal{H})_+.$$

Then $\mathrm{Tr}' = \mathrm{Tr}$.

Proof

There exists a unitary $u \in \mathsf{B}(\mathcal{H})$ such that $u\eta_j = \xi_j$ for all $j \in J$. Therefore

$$\mathrm{Tr}'(t) = \sum_{j\in J} \langle u^*\xi_j | tu^*\xi_j \rangle = \sum_{j\in J} \langle \xi_j | (utu^*)\xi_j \rangle = \mathrm{Tr}(utu^*) = \mathrm{Tr}(t)$$

for any $t \in \mathsf{B}(\mathcal{H})_+$. □

Corollary 6.3 shows that the function

$$\mathrm{Tr} : \mathsf{B}(\mathcal{H})_+ \ni t \longmapsto \sum_{i\in I} \langle \xi_i | t\xi_i \rangle \in [0, +\infty]$$

does not depend on the choice of the orthonormal basis $\{\xi_i\}_{i\in I}$ of $\mathcal{H}$. We call this function the *trace*.

The following properties of the trace are immediate from its definition:

- for $t \in \mathsf{B}(\mathcal{H})_+$ and $\lambda \in \mathbb{R}_+$ we have $\mathrm{Tr}(\lambda t) = \lambda\,\mathrm{Tr}(t)$,
- if $t, r \in \mathsf{B}(\mathcal{H})_+$ and $t \geq r$ then $\mathrm{Tr}(t) \geq \mathrm{Tr}(r)$,
- for $t, r \in \mathsf{B}(\mathcal{H})_+$ we have $\mathrm{Tr}(t + r) = \mathrm{Tr}(t) + \mathrm{Tr}(r)$,

with the last property following from standard facts about series with positive terms.

Proposition 6.4 *For $t \in \mathsf{B}(\mathcal{H})_+$ we have* $\mathrm{Tr}(t) \geq \|t\|$.

Proof
Put $r = t^{\frac{1}{2}}$. We have $\|t\| = \|r^*r\| = \|r\|^2$. Take $\varepsilon > 0$ and let $\psi \in \mathcal{H}$ be such that $\|\psi\| = 1$ and

$$\|r\psi\| > \|r\| - \varepsilon.$$

Then

$$\langle \psi | t\psi \rangle = \|r\psi\|^2 > \|r\|^2 - 2\varepsilon\|r\| + \varepsilon^2,$$

so that

$$\langle \psi | t\psi \rangle > \|t\| - 2\varepsilon\|t\|^{\frac{1}{2}} + \varepsilon^2.$$

Now let us choose an orthonormal basis $\{\xi_j\}_{j\in J}$ of $\mathcal{H}$ such that $\psi = \xi_{j_0}$ for some $j_0 \in J$. Then

$$\mathrm{Tr}(t) \geq \langle \psi | t\psi \rangle > \|t\| - 2\varepsilon\|t\|^{\frac{1}{2}} + \varepsilon^2.$$

As $\varepsilon > 0$ is arbitrary, we obtain $\mathrm{Tr}(t) \geq \|t\|$. □

6.2 Trace Class and Hilbert-Schmidt Operators

Define

$$\begin{aligned} \mathsf{B}_1(\mathcal{H}) &= \mathrm{span}\,\big\{t \in \mathsf{B}(\mathcal{H})_+ \,\big|\, \mathrm{Tr}(t) < +\infty\big\}, \\ \mathsf{B}_2(\mathcal{H}) &= \big\{t \in \mathsf{B}(\mathcal{H}) \,\big|\, \mathrm{Tr}(t^*t) < +\infty\big\}. \end{aligned} \tag{6.2}$$

Elements of the subspace $\mathsf{B}_1(\mathcal{H}) \subset \mathsf{B}(\mathcal{H})$ are called the *trace class* operators, while $\mathsf{B}_2(\mathcal{H})$ is the set of *Hilbert-Schmidt operators*.

Before proceeding with analysis of the classes $\mathsf{B}_1(\mathcal{H})$ and $\mathsf{B}_2(\mathcal{H})$ let us note that for any $a, b \in \mathsf{B}(\mathcal{H})$ we have

$$ab = \tfrac{1}{4}\sum_{k=0}^{3} \mathrm{i}^k (b + \mathrm{i}^k \mathbb{1})^* a (b + \mathrm{i}^k \mathbb{1}). \tag{6.3}$$

Indeed:

$$\begin{aligned}
\sum_{k=0}^{3} \mathrm{i}^k (b + \mathrm{i}^k \mathbb{1})^* a (b + \mathrm{i}^k \mathbb{1}) = (b + \mathbb{1})^* a (b + \mathbb{1})& \\
+ \mathrm{i}(b + \mathrm{i}\mathbb{1})^* a (b + \mathrm{i}\mathbb{1})& \\
- (b - \mathbb{1})^* a (b - \mathbb{1})& \\
- \mathrm{i}(b - \mathrm{i}\mathbb{1})^* a (b - \mathrm{i}\mathbb{1})& \\
= b^*ab + b^*a + ab + a& \\
+ \mathrm{i}b^*ab + ab + \mathrm{i}b^*a + \mathrm{i}a& \\
- b^*ab + ab - b^*a - a& \\
- \mathrm{i}b^*ab - \mathrm{i}b^*a + ab - \mathrm{i}a = 4ab.&
\end{aligned}$$

Lemma 6.5 *Let $x \in \mathsf{B}(\mathcal{H})$ be such that* $\mathrm{Tr}\left(|x|^p\right) < +\infty$ *for some $p > 0$. Then $x \in \mathsf{B}_0(\mathcal{H})$.*

Proof

Let $\{\psi_j\}_{j \in J}$ be an orthonormal basis of $\mathcal{H}$. For any $\varepsilon > 0$ there exists a finite subset $J_\varepsilon \subset J$ such that

$$\sum_{j \notin J_\varepsilon} \langle \psi_j \,|\, |x|^p \psi_j \rangle < \varepsilon.$$

Let p_ε be the projection onto $\mathrm{span}\{\psi_j \mid j \in J_\varepsilon\}$. Then by Proposition 6.4 we have

$$\begin{aligned}
\left\| |x|^{\frac{p}{2}} - |x|^{\frac{p}{2}} p_\varepsilon \right\|^2 &= \left\| |x|^{\frac{p}{2}} (\mathbb{1} - p_\varepsilon) \right\|^2 \\
&= \left\| (\mathbb{1} - p_\varepsilon) |x|^p (\mathbb{1} - p_\varepsilon) \right\| \\
&\leq \mathrm{Tr}\left((\mathbb{1} - p_\varepsilon) |x|^p (\mathbb{1} - p_\varepsilon) \right) \\
&= \sum_{j \notin J_\varepsilon} \langle \psi_j \,|\, |x|^p \psi_j \rangle < \varepsilon.
\end{aligned}$$

The operator $|x|^{\frac{p}{2}} p_\varepsilon$ is finite dimensional, and consequently $|x|^{\frac{p}{2}}$ is compact. Now since $|x| = f\big(|x|^{\frac{p}{2}}\big)$, where $f(\lambda) = \lambda^{\frac{2}{p}}$ for all $\lambda \in \sigma\big(|x|\big)$, it follows that also $|x| \in \mathsf{B}_0(\mathcal{H})$ and finally, writing x in its polar decomposition, we obtain $x = u|x| \in \mathsf{B}_0(\mathcal{H})$. □

Application of Lemma 6.5 to the case $p = 2$ immediately yields the third inclusion of the next theorem:

Theorem 6.6
We have

$$\mathcal{F}(\mathcal{H}) \subset \mathsf{B}_1(\mathcal{H}) \subset \mathsf{B}_2(\mathcal{H}) \subset \mathsf{B}_0(\mathcal{H}) \tag{6.4}$$

and each of these subsets is a self-adjoint ideal in $\mathsf{B}(\mathcal{H})$. *Moreover*

$$\mathsf{B}_1(\mathcal{H}) = \big\{x \in \mathsf{B}(\mathcal{H}) \,\big|\, \mathrm{Tr}\left(|x|\right) < +\infty\big\}. \tag{6.5}$$

Proof
We begin by noticing that all sets in (6.4) are closed under taking adjoints. In ▶ Sect. 5.1 we indicated that $\mathcal{F}(\mathcal{H})$ and $\mathsf{B}_0(\mathcal{H})$ are ideals in $\mathsf{B}(\mathcal{H})$.

We will now show that $\mathsf{B}_1(\mathcal{H})$ is a right ideal in $\mathsf{B}(\mathcal{H})$. Take a positive $a \in \mathsf{B}(\mathcal{H})$ such that $\mathrm{Tr}(a) < +\infty$ and let $b \in \mathsf{B}(\mathcal{H})$. From (6.3) we know that

$$ab = \tfrac{1}{4}\sum_{k=0}^{3} \mathrm{i}^k {v_k}^* a v_k,$$

where $v_k = (b + \mathrm{i}^k \mathbb{1})$. For each k we have

$$\begin{aligned}\mathrm{Tr}({v_k}^* a v_k) &= \mathrm{Tr}\left({v_k}^* a^{\frac{1}{2}} a^{\frac{1}{2}} v_k\right) = \mathrm{Tr}\left((a^{\frac{1}{2}} v_k)^* (a^{\frac{1}{2}} v_k)\right) \\ &= \mathrm{Tr}\left((a^{\frac{1}{2}} v_k)(a^{\frac{1}{2}} v_k)^*\right) = \mathrm{Tr}(a^{\frac{1}{2}} v_k {v_k}^* a^{\frac{1}{2}}) \\ &\le \mathrm{Tr}\left(\|v_k {v_k}^*\| a\right) = \|v_k\|^2\, \mathrm{Tr}(a) < +\infty,\end{aligned}$$

and hence $ab \in \mathsf{B}_1(\mathcal{H})$. Since the space $\mathsf{B}_1(\mathcal{H})$ is spanned by a as above, it is a right ideal in $\mathsf{B}(\mathcal{H})$. As a self-adjoint subset of $\mathsf{B}(\mathcal{H})$ it must also be a left ideal.[1]

This immediately shows that $\mathsf{B}_1(\mathcal{H})$ contains the set

$$\big\{x \in \mathsf{B}(\mathcal{H}) \,\big|\, \mathrm{Tr}\left(|x|\right) < +\infty\big\}.$$

[1] Any right ideal J in $\mathsf{B}(\mathcal{H})$ closed under taking adjoints is also a left ideal: if $x \in \mathsf{J}$ and $y \in \mathsf{B}(\mathcal{H})$ then $yx = (x^* y^*)^* \in \mathsf{J}$.

Indeed: if $x \in \mathsf{B}(\mathcal{H})$ and $\operatorname{Tr}\left(|x|\right) < +\infty$ then clearly $|x| \in \mathsf{B}_1(\mathcal{H})$ and writing x as $x = u|x|$ (polar decomposition) we find that $x \in \mathsf{B}_1(\mathcal{H})$, as $\mathsf{B}_1(\mathcal{H})$ is an ideal.

Conversely, if $t \in \mathsf{B}_1(\mathcal{H})$ then also $|t| = u^*t$ belongs to $\mathsf{B}_1(\mathcal{H})$ (here, again, $t = u|t|$ is the polar decomposition). This means that $|t|$ can be written in the form

$$|t| = \sum_{i=1}^{n} \alpha_i d_i$$

for some $\alpha_1, \dots, \alpha_n \in \mathbb{C}$ and positive operators $d_1, \dots, d_n$ with finite trace. Since, as is easily checked,

$$|t| \le \sum_{i=1}^{n} |\alpha_i| d_i,$$

we have $\operatorname{Tr}\left(|t|\right) \le \sum_{i=1}^{n} |\alpha_i| \operatorname{Tr}(d_i) < +\infty$, i.e. $t \in \left\{x \in \mathsf{B}(\mathcal{H}) \,\middle|\, \operatorname{Tr}\left(|x|\right) < +\infty\right\}$. This way we proved (6.5).

We will now deal with the set $\mathsf{B}_2(\mathcal{H})$. It follows from the identity

$$(a+b)^*(a+b) + (a-b)^*(a-b) = 2(a^*a + b^*b), \qquad a, b \in \mathsf{B}(\mathcal{H})$$

that $(a+b)^*(a+b) \le 2(a^*a + b^*b)$, and thus $\mathsf{B}_2(\mathcal{H})$ is a vector subspace of $\mathsf{B}(\mathcal{H})$. $\mathsf{B}_2(\mathcal{H})$ is, moreover, a left ideal, since if $t \in \mathsf{B}_2(\mathcal{H})$ then for any $s \in \mathsf{B}(\mathcal{H})$ we have

$$(st)^*(st) = t^*s^*st \le \|s\|^2 t^*t,$$

so that $\operatorname{Tr}\left((st)^*(st)\right) \le \|s\|^2 \operatorname{Tr}(t^*t) < +\infty$, i.e. $st \in \mathsf{B}_2(\mathcal{H})$. As $\mathsf{B}_2(\mathcal{H})$ is a self-adjoint subset of $\mathsf{B}(\mathcal{H})$, it is also a right ideal.

Now let $x \in \mathcal{F}(\mathcal{H})$. Then $|x|$ is a positive finite dimensional operator.[2] Therefore $|x|$ has finite trace and consequently $|x| \in \mathsf{B}_1(\mathcal{H})$. The latter set is an ideal, so $x = u|x| \in \mathsf{B}_1(\mathcal{H})$ and we have proved that

$$\mathcal{F}(\mathcal{H}) \subset \mathsf{B}_1(\mathcal{H}).$$

To finish the proof we only need to show that $\mathsf{B}_1(\mathcal{H}) \subset \mathsf{B}_2(\mathcal{H})$. Let $x \in \mathsf{B}_1(\mathcal{H})$. We have

$$x^*x = |x|^2 = |x|^{\frac{1}{2}}\left(|x|^{\frac{1}{2}}\right)^*\left(|x|^{\frac{1}{2}}\right)|x|^{\frac{1}{2}} \le \|x\| |x|^{\frac{1}{2}} |x|^{\frac{1}{2}} = \|x\| |x|.$$

Thus $\operatorname{Tr}(x^*x) \le \|x\| \operatorname{Tr}\left(|x|\right) < +\infty$, i.e. $\mathsf{B}_1(\mathcal{H}) \subset \mathsf{B}_2(\mathcal{H})$. □

---

[2] Let us write x in the form $x = \sum_{i=1}^{N} |\psi_i\rangle\langle\varphi_i|$, with $\{\psi_1, \dots, \psi_N\}$ orthonormal. Then $x^*x = \sum_{i=1}^{N} |\varphi_i\rangle\langle\varphi_i|$ is a positive operator on $\operatorname{span}\{\varphi_1, \dots, \varphi_N\}$. Let $\{\xi_1, \dots, \xi_M\}$ be an orthonormal basis of $\operatorname{span}\{\varphi_1, \dots, \varphi_N\}$ consisting of eigenvectors of x^*x. Then $x^*x = \sum_{j=1}^{M} \lambda_j |\xi_j\rangle\langle\xi_j|$ and by uniqueness of square roots we have $|x| = \sum_{j=1}^{M} \sqrt{\lambda_j}\, |\xi_j\rangle\langle\xi_j|$.

Remark 6.7 We have $\{t \in \mathsf{B}(\mathcal{H})_+ \mid \mathrm{Tr}(t) < \infty\} = \mathsf{B}_1(\mathcal{H}) \cap \mathsf{B}(\mathcal{H})_+$. Indeed: it follows from the definition (6.2) that

$$\{t \in \mathsf{B}(\mathcal{H})_+ \mid \mathrm{Tr}(t) < \infty\} \subset \mathsf{B}_1(\mathcal{H}) \cap \mathsf{B}(\mathcal{H})_+.$$

On the other hand, if $a \in \mathsf{B}_1(\mathcal{H}) \cap \mathsf{B}(\mathcal{H})_+$ then $a = |a|$ and

$$a \in \{x \in \mathsf{B}(\mathcal{H}) \mid \mathrm{Tr}(|x|) < +\infty\}.$$

Therefore $\{t \in \mathsf{B}(\mathcal{H})_+ \mid \mathrm{Tr}(t) < \infty\} \supset \mathsf{B}_1(\mathcal{H}) \cap \mathsf{B}(\mathcal{H})_+$.

Proposition 6.8 *The function* $\mathsf{B}_1(\mathcal{H}) \cap \mathsf{B}(\mathcal{H})_+ \ni x \mapsto \mathrm{Tr}(x) \in \mathbb{R}_+$ *extends uniquely to a linear functional on* $\mathsf{B}_1(\mathcal{H})$.

Proof

Since $\mathsf{B}_1(\mathcal{H}) = \mathrm{span}\,(\mathsf{B}_1(\mathcal{H}) \cap \mathsf{B}(\mathcal{H})_+)$, it is enough to prove that an extension of Tr to $\mathsf{B}_1(\mathcal{H})$ exists. We will show that the formula

$$\sum_{i=1}^{N} \alpha_i x_i \longmapsto \sum_{i=1}^{N} \alpha_i \,\mathrm{Tr}(x_i)$$

provides one. To see this it is enough to check that if $\sum_{i=1}^{N} \alpha_i x_i = 0$ (for some $\alpha_1, \ldots, \alpha_N \in \mathbb{C}$ and $x_1, \ldots, x_N \in \mathsf{B}(\mathcal{H})_+$ with finite trace), then $\sum_{i=1}^{N} \alpha_i \,\mathrm{Tr}(x_i) = 0$.

Indeed: in this case we have

$$\sum_{i=1}^{N} \mathrm{Re}\,\alpha_i\, x_i = 0 \quad \text{and} \quad \sum_{i=1}^{N} \mathrm{Im}\,\alpha_i\, x_i = 0.$$

Let $\{1, \ldots, N\} = A \cup B$ with A consisting of those i for which $\mathrm{Re}\,\alpha_i \geq 0$ and B of the remaining elements of $\{1, \ldots, N\}$ (so that $\mathrm{Re}\,\alpha_i < 0$ for $i \in B$). Then

$$\sum_{i \in A} \mathrm{Re}\,\alpha_i\, x_i = \sum_{i \in B} (-\mathrm{Re}\,\alpha_i) x_i$$

are linear combinations of positive operators with positive coefficients. Thus from additivity of the trace we obtain

$$\sum_{i \in A} \mathrm{Re}\,\alpha_i\, \mathrm{Tr}(x_i) = \sum_{i \in B} (-\mathrm{Re}\,\alpha_i)\, \mathrm{Tr}(x_i),$$

i.e.

$$\sum_{i=1}^{N} \mathrm{Re}\,\alpha_i\, \mathrm{Tr}(x_i) = 0.$$

Similarly we get $\sum_{i=1}^{N} \operatorname{Im} \alpha_i \operatorname{Tr}(x_i) = 0$. □

Let $x \in \mathsf{B}_1(\mathcal{H})$. The value of the linear extension of Tr from $\mathsf{B}_1(\mathcal{H}) \cap \mathsf{B}(\mathcal{H})_+$ to $\mathsf{B}_1(\mathcal{H})$ on x will also be called the *trace* of x and just like for positive operators, we will also denote it by $\operatorname{Tr}(x)$. We will now show that the value of $\operatorname{Tr}(x)$ can be calculated by the same formula as for positive operators.

Take any $x \in \mathsf{B}_1(\mathcal{H})$ and let $\{\xi_i\}_{i\in I}$ be an orthonormal basis of $\mathcal{H}$. Using the polar decomposition of x write $x = u|x| = u|x|^{\frac{1}{2}}|x|^{\frac{1}{2}}$. Then

$$\left| \langle \xi_i | x\xi_i \rangle \right| = \left| \left\langle |x|^{\frac{1}{2}} u^* \xi_i \middle| |x|^{\frac{1}{2}} \xi_i \right\rangle \right| \leq \left\| |x|^{\frac{1}{2}} u^* \xi_i \right\| \left\| |x|^{\frac{1}{2}} \xi_i \right\|.$$

Now, since

$$\sum_{i\in I} \left\| |x|^{\frac{1}{2}} \xi_i \right\|^2 = \operatorname{Tr}\left(|x|\right) < +\infty$$

and

$$\sum_{i\in I} \left\| |x|^{\frac{1}{2}} u^* \xi_i \right\|^2 = \operatorname{Tr}\left(u|x|u^*\right) < +\infty$$

(i.e. $|x|^{\frac{1}{2}}$ and $|x|^{\frac{1}{2}}u^*$ belong to $\mathsf{B}_2(\mathcal{H})$), we obtain

$$\begin{aligned}\sum_{i\in I} \left| \langle \xi_i | x\xi_i \rangle \right| &\leq \sum_{i\in I} \| |x|^{\frac{1}{2}} u^* \xi_i \| \| |x|^{\frac{1}{2}} \xi_i \| \\ &\leq \left(\sum_{i\in I} \| |x|^{\frac{1}{2}} u^* \xi_i \|^2 \right)^{\frac{1}{2}} \left(\sum_{i\in I} \| |x|^{\frac{1}{2}} \xi_i \|^2 \right)^{\frac{1}{2}} < +\infty.\end{aligned}$$

It follows that the series $\sum_{i\in I} \langle \xi_i | x\xi_i \rangle$ is absolutely convergent. Furthermore, because x can be expressed as $x = \sum_{k=1}^{N} \alpha_k x_k$, where $\alpha_1, \ldots, \alpha_N \in \mathbb{C}$ and the operators $x_1, \ldots, x_N$ are positive with finite trace, we have

$$\begin{aligned}\sum_{i\in I} \langle \xi_i | x\xi_i \rangle &= \sum_{i\in I} \left\langle \xi_i \middle| \sum_{k=1}^{N} \alpha_k x_k \xi_i \right\rangle \\ &= \sum_{i\in I} \sum_{k=1}^{N} \alpha_k \langle \xi_i | x_k \xi_i \rangle = \sum_{k=1}^{N} \alpha_k \sum_{i\in I} \langle \xi_i | x_k \xi_i \rangle,\end{aligned}$$

and the last expression is independent of the choice of the basis $\{\xi_i\}_{i\in I}$. This way we have established the following:

Proposition 6.9 *Let* $x \in \mathsf{B}_1(\mathcal{H})$ *and let* $\{\xi_i\}_{i\in I}$ *be an orthonormal basis of* $\mathcal{H}$. *Then the series*

$$\sum_{i\in I} \langle \xi_i \,|\, x\xi_i \rangle \tag{6.6}$$

is absolutely convergent and its sum is independent of the choice of the basis $\{\xi_i\}_{i\in I}$.

Since (6.6) is a linear functional on $\mathsf{B}_1(\mathcal{H})$ which coincides with $\mathrm{Tr}(\cdot)$ on $\mathsf{B}_1(\mathcal{H}) \cap \mathsf{B}(\mathcal{H})_+$, we see that

$$\mathrm{Tr}(x) = \sum_{i\in I} \langle \xi_i \,|\, x\xi_i \rangle$$

for any $x \in \mathsf{B}_1(\mathcal{H})$ and any orthonormal basis $\{\xi_i\}_{i\in I}$ of $\mathcal{H}$.

Remark 6.10 It is worth remembering that if $x \in \mathsf{B}(\mathcal{H})$ and for *some* orthonormal basis $\{\xi_i\}_{i\in I}$ of $\mathcal{H}$ the series (6.6) converges (absolutely) it does not necessarily mean that $x \in \mathsf{B}_1(\mathcal{H})$. For example let $\mathcal{H}$ be infinite dimensional, choose an orthonormal basis $\{\xi_i\}_{i\in I}$ of $\mathcal{H}$ and take u to be the unitary operator such that $u\xi_i = \xi_{\pi(i)}$, where π is a permutation of the set I with finite number of fixed points. Then the series $\sum_{i\in I} \langle \xi_i \,|\, u\xi_i \rangle$ is absolutely convergent, but $|u| = \mathbb{1}$ has infinite trace.

Consider now $x, y \in \mathsf{B}_2(\mathcal{H})$. In a way similar to the proof of formula (6.3), we can prove the following analog of the polarization identity:

$$x^*y = \frac{1}{4}\sum_{k=0}^{3} \mathrm{i}^k (y + \mathrm{i}^k x)^*(y + \mathrm{i}^k x). \tag{6.7}$$

As $\mathsf{B}_2(\mathcal{H})$ is a vector subspace of $\mathsf{B}(\mathcal{H})$, we have $y + \mathrm{i}^k x \in \mathsf{B}_2(\mathcal{H})$, and consequently $\mathrm{Tr}\left((y + \mathrm{i}^k x)^*(y + \mathrm{i}^k x)\right) < +\infty$ for each k. In particular x^*y belongs to $\mathsf{B}_1(\mathcal{H})$.

Theorem 6.11
The ideal $\mathsf{B}_2(\mathcal{H})$ *is a Hilbert space with scalar product*

$$\langle x \,|\, y \rangle_{\mathrm{Tr}} = \mathrm{Tr}(x^*y), \qquad x, y \in \mathsf{B}_2(\mathcal{H}). \tag{6.8}$$

Proof
We already know that for $x, y \in \mathsf{B}_2(\mathcal{H})$ we have $x^*y \in \mathsf{B}_1(\mathcal{H})$. It follows that formula (6.8) makes sense and defines a sesquilinear form (linear in y and anti-linear in x) on $\mathsf{B}_2(\mathcal{H})$. This form is hermitian, as

$$\mathrm{Tr}(y^*x) = \sum_{i\in I} \langle \psi_i | y^* x \psi_i \rangle = \sum_{i\in I} \langle x^* y \psi_i | \psi_i \rangle = \overline{\sum_{i\in I} \langle \psi_i | x^* y \psi_i \rangle} = \overline{\mathrm{Tr}(x^*y)}$$

for any orthonormal basis $\{\psi_i\}_{i\in I}$ of $\mathcal{H}$. Moreover, it is strictly positive, since

$$\langle x | x \rangle_{\mathrm{Tr}} = \mathrm{Tr}(x^*x) \geq \|x^*x\| = \|x\|^2. \tag{6.9}$$

Let $\|\cdot\|_2$ denote the norm on $\mathsf{B}_2(\mathcal{H})$ associated with $\langle\cdot|\cdot\rangle_{\mathrm{Tr}}$. It remains to prove that $\big(\mathsf{B}_2(\mathcal{H}), \|\cdot\|_2\big)$ is a complete space. First note that if $(x_n)_{n\in\mathbb{N}}$ is a Cauchy sequence in $\mathsf{B}_2(\mathcal{H})$ then the inequality (6.9) shows that $(x_n)_{n\in\mathbb{N}}$ is also a Cauchy sequence for the operator norm. Let x be the limit of $(x_n)_{n\in\mathbb{N}}$ in $\mathsf{B}(\mathcal{H})$.

In order to estimate $\|x - x_n\|_2$ let us fix an orthonormal basis $\{\psi_i\}_{i\in I}$ of $\mathcal{H}$. Now $x_m \xrightarrow[m\to\infty]{} x$ in norm and therefore also in strong topology. Therefore, for a finite subset $I_0 \subset I$ we have

$$\begin{aligned}\sum_{i\in I_0} \big\|(x - x_n)\psi_i\big\|^2 &= \lim_{m\to\infty} \sum_{i\in I_0} \big\|(x_m - x_n)\psi_i\big\|^2 \\ &\leq \limsup_{m\to\infty} \sum_{i\in I} \big\|(x_m - x_n)\psi_i\big\|^2 = \limsup_{m\to\infty} \|x_m - x_n\|_2^2.\end{aligned}$$

It follows that

$$\|x - x_n\|_2^2 = \sup_{\substack{I_0\subset I\\ |I_0|<+\infty}} \sum_{i\in I_0} \big\|(x - x_n)\psi_i\big\|^2 \leq \limsup_{m\to\infty} \|x_m - x_n\|_2^2.$$

In particular $x \in \mathsf{B}_2(\mathcal{H})$ and x is the limit of $(x_n)_{n\in\mathbb{N}}$ in the norm $\|\cdot\|_2$. □

The norm

$$\|x\|_2 = \big(\mathrm{Tr}(x^*x)\big)^{\frac{1}{2}}, \qquad x \in \mathsf{B}_2(\mathcal{H})$$

introduced in the proof of Theorem 6.11 is called the *Hilbert-Schmidt norm.*

Proposition 6.12 *Let $\{\psi_i\}_{i\in I}$ be an orthonormal basis of $\mathcal{H}$. Then the system of operators*

$$\big\{\, |\psi_i\rangle\langle\psi_j| \,\big\}_{i,j\in I} \tag{6.10}$$

is an orthonormal basis of the Hilbert space $\mathsf{B}_2(\mathcal{H})$.

Proof
We easily check that the system (6.10) is orthonormal. Furthermore, for $x \in \mathsf{B}_2(\mathcal{H})$ we have

$$\begin{aligned}\big\langle\, |\psi_i\rangle\langle\psi_j|\,\big|x\big\rangle_{\mathrm{Tr}} &= \mathrm{Tr}\left(\,|\psi_j\rangle\langle\psi_i|\,x\right) = \sum_{k\in I}\Big\langle\psi_k\Big|\,|\psi_j\rangle\langle\psi_i|\,x\psi_k\Big\rangle \\ &= \sum_{k\in I}\Big\langle\,|\psi_i\rangle\langle\psi_j|\,\psi_k\Big|x\psi_k\Big\rangle = \langle\psi_i|x\psi_j\rangle.\end{aligned}$$

Thus, if x is orthogonal to all elements of (6.10) then $x = 0$. □

Theorem 6.13
Assume that either
(a) $x, y \in \mathsf{B}_2(\mathcal{H})$.
or
(b) $x \in \mathsf{B}_1(\mathcal{H})$ *and* $y \in \mathsf{B}(\mathcal{H})$
Then

$$\mathrm{Tr}(xy) = \mathrm{Tr}(yx).$$

Proof
To prove the theorem under assumption (a) we will use formula (6.7). We have

$$\begin{aligned}\mathrm{Tr}(x^*y) &= \tfrac{1}{4}\sum_{k=0}^{3}\mathrm{i}^k\,\mathrm{Tr}\left((y+\mathrm{i}^kx)^*(y+\mathrm{i}^kx)\right) \\ &= \tfrac{1}{4}\sum_{k=0}^{3}\mathrm{i}^k\,\mathrm{Tr}\left((y+\mathrm{i}^kx)(y+\mathrm{i}^kx)^*\right) \\ &= \tfrac{1}{4}\sum_{k=0}^{3}\mathrm{i}^k\,\mathrm{Tr}\left((y^*-\mathrm{i}^kx^*)^*(y^*-\mathrm{i}^kx^*)\right) \\ &= \tfrac{1}{4}\sum_{k=0}^{3}\mathrm{i}^k\,\mathrm{Tr}\left(\left((-\mathrm{i}^k)(\mathrm{i}^ky^*+x^*)\right)^*(-\mathrm{i}^k)(\mathrm{i}^ky^*+x^*)\right) \\ &= \tfrac{1}{4}\sum_{k=0}^{3}\mathrm{i}^k\,\mathrm{Tr}\left(\mathrm{i}^k(\mathrm{i}^ky^*+x^*)^*(-\mathrm{i}^k)(\mathrm{i}^ky^*+x^*)\right)\end{aligned}$$

$$= \tfrac{1}{4}\sum_{k=0}^{3} \mathrm{i}^k \operatorname{Tr}\left((\mathrm{i}^k y^* + x^*)^*(\mathrm{i}^k y^* + x^*)\right)$$

$$= \tfrac{1}{4}\sum_{k=0}^{3} \mathrm{i}^k \operatorname{Tr}\left((x^* + \mathrm{i}^k y^*)^*(x^* + \mathrm{i}^k y^*)\right) = \operatorname{Tr}(yx^*)$$

and since $\mathsf{B}_2(\mathcal{H})$ is closed under involution of $\mathsf{B}(\mathcal{H})$, we obtain $\operatorname{Tr}(xy) = \operatorname{Tr}(yx)$ for all $x, y \in \mathsf{B}_2(\mathcal{H})$.

On the other hand, with assumption (b) we can write x as a linear combination of positive operators with finite trace. If z is such an operator then by the first part of the theorem we have

$$\operatorname{Tr}(zy) = \operatorname{Tr}\left(z^{\frac{1}{2}}(z^{\frac{1}{2}}y)\right) = \operatorname{Tr}\left((z^{\frac{1}{2}}y)z^{\frac{1}{2}}\right)$$
$$= \operatorname{Tr}\left(z^{\frac{1}{2}}(yz^{\frac{1}{2}})\right) = \operatorname{Tr}\left((yz^{\frac{1}{2}})z^{\frac{1}{2}}\right) = \operatorname{Tr}(yz),$$

because $z^{\frac{1}{2}}$, $z^{\frac{1}{2}}y$ and $yz^{\frac{1}{2}}$ belong to $\mathsf{B}_2(\mathcal{H})$. □

Lemma 6.14 *Let* $x \in \mathsf{B}_1(\mathcal{H})$ *and* $y \in \mathsf{B}(\mathcal{H})$. *Then*

$$\left|\operatorname{Tr}(yx)\right| \le \|y\| \operatorname{Tr}\left(|x|\right).$$

Proof

Let $x = u|x|$ be the polar decomposition of x. Since $\operatorname{Tr}\left(|x|\right) < +\infty$, we see that $|x|^{\frac{1}{2}} \in \mathsf{B}_2(\mathcal{H})$. Now the Schwarz inequality (for the scalar product $\langle\cdot|\cdot\rangle_{\mathrm{Tr}}$) gives

$$\left|\operatorname{Tr}(yx)\right|^2 = \left|\operatorname{Tr}(yu|x|^{\frac{1}{2}}|x|^{\frac{1}{2}})\right|^2$$
$$= \left|\left\langle |x|^{\frac{1}{2}}u^*y^* \,\middle|\, |x|^{\frac{1}{2}}\right\rangle_{\mathrm{Tr}}\right|$$
$$\le \left\||x|^{\frac{1}{2}}u^*y^*\right\|_2^2 \left\||x|^{\frac{1}{2}}\right\|_2^2$$
$$= \operatorname{Tr}\left(yu|x|^{\frac{1}{2}}|x|^{\frac{1}{2}}u^*y^*\right)\operatorname{Tr}\left(|x|\right)$$
$$= \operatorname{Tr}\left(|x|^{\frac{1}{2}}u^*y^*yu|x|^{\frac{1}{2}}\right)\operatorname{Tr}\left(|x|\right)$$
$$\le \operatorname{Tr}\left(\|yu\|^2|x|^{\frac{1}{2}}|x|^{\frac{1}{2}}\right)\operatorname{Tr}\left(|x|\right) \le \|y\|^2 \operatorname{Tr}\left(|x|\right)^2.$$

□

Theorem 6.15
The ideal $\mathsf{B}_1(\mathcal{H}) \subset \mathsf{B}(\mathcal{H})$ *is a Banach algebra with norm*

$$\|x\|_1 = \operatorname{Tr}\big(|x|\big), \qquad x \in \mathsf{B}_1(\mathcal{H}).$$

Proof
Obviously we have $\|\alpha x\|_1 = |\alpha|\|x\|_1$ for all $x \in \mathsf{B}_1(\mathcal{H})$ and $\alpha \in \mathbb{C}$. Moreover, since

$$\|x\|_1 = \operatorname{Tr}\big(|x|\big) \geq \big\||x|\big\| = \|x\|,$$

we see that $\|x\|_1 = 0$ implies $x = 0$.

Take $x, y \in \mathsf{B}_1(\mathcal{H})$ and let $x + y = v|x + y|$ be the polar decomposition of $x + y$. Then

$$\begin{aligned}\|x + y\|_1 &= \operatorname{Tr}\big(|x + y|\big) = \operatorname{Tr}\big(v^*(x + y)\big)\\ &= \operatorname{Tr}(v^*x) + \operatorname{Tr}(v^*y) = \big|\operatorname{Tr}(v^*x) + \operatorname{Tr}(v^*y)\big|\\ &\leq \big|\operatorname{Tr}(v^*x)\big| + \big|\operatorname{Tr}(v^*y)\big| \leq \|v^*\| \operatorname{Tr}\big(|x|\big) + \|v^*\| \operatorname{Tr}\big(|y|\big)\\ &\leq \|x\|_1 + \|y\|_1\end{aligned}$$

by Lemma 6.14. This shows that $\|\cdot\|_1$ is a norm on $\mathsf{B}_1(\mathcal{H})$.

Using a similar technique we prove submultiplicativity of $\|\cdot\|_1$: take $x, y \in \mathsf{B}_1(\mathcal{H})$ and write the polar decomposition of xy as $xy = u|xy|$. We have

$$\begin{aligned}\|xy\|_1 = \operatorname{Tr}\big(u^*xy\big) = \big|\operatorname{Tr}(u^*xy)\big| &\leq \|u^*x\| \operatorname{Tr}\big(|y|\big)\\ &\leq \|x\| \operatorname{Tr}\big(|y|\big) \leq \|x\|_1 \operatorname{Tr}\big(|y|\big) = \|x\|_1\|y\|_1.\end{aligned}$$

Completeness of $\mathsf{B}_1(\mathcal{H})$ follows from the next theorem. □

Theorem 6.16
For $x \in \mathsf{B}_1(\mathcal{H})$ *define*

$$\varphi_x : \mathsf{B}_0(\mathcal{H}) \ni y \longmapsto \operatorname{Tr}(xy) \in \mathbb{C}.$$

Then $\varphi_x \in \mathsf{B}_0(\mathcal{H})^*$ *and* $x \mapsto \varphi_x$ *is an isometry of* $\mathsf{B}_1(\mathcal{H})$ *with the norm* $\|\cdot\|_1$ *onto* $\mathsf{B}_0(\mathcal{H})^*$.

Proof

For $x \in \mathsf{B}_1(\mathcal{H})$ and $y \in \mathsf{B}_0(\mathcal{H})$ we have

$$\left|\varphi_x(y)\right| = \left|\operatorname{Tr}(xy)\right| = \left|\operatorname{Tr}(yx)\right| \le \|y\| \operatorname{Tr}\left(|x|\right) = \|y\|\|x\|_1,$$

so that $\varphi_x \in \mathsf{B}_0(\mathcal{H})^*$ and $\|\varphi_x\| \le \|x\|_1$.

Take $\varphi \in \mathsf{B}_0(\mathcal{H})^*$. Then for $s \in \mathsf{B}_2(\mathcal{H}) \subset \mathsf{B}_0(\mathcal{H})$ we have

$$\left|\varphi(s)\right| \le \|\varphi\|\|s\| \le \|\varphi\|\|s\|_2,$$

and hence φ is a continuous linear functional on the Hilbert space $\mathsf{B}_2(\mathcal{H})$. Thus there exists $x \in \mathsf{B}_2(\mathcal{H})$ such that

$$\varphi(s) = \left\langle x^* \middle| s \right\rangle_{\operatorname{Tr}}, \qquad s \in \mathsf{B}_2(\mathcal{H}).$$

Let $x = v|x|$ be the polar decomposition of x. Take an orthonormal basis $\{\psi_j\}_{j\in J}$ of $\mathcal{H}$ and for a finite $J_0 \subset J$ let p_0 be the projection onto $\operatorname{span}\{\psi_j \mid j \in J_0\}$. Then, using the fact that $x, p_0 \in \mathsf{B}_2(\mathcal{H})$ and Theorem 6.13, we compute

$$\begin{aligned}\sum_{j\in J_0} \left\langle \psi_j \middle| |x| \psi_j \right\rangle &= \left| \sum_{j\in J_0} \left\langle \psi_j \middle| |x| \psi_j \right\rangle \right| = \left|\operatorname{Tr}(p_0|x|)\right| = \left|\operatorname{Tr}(p_0 v^* x)\right| \\ &= \left|\operatorname{Tr}(x p_0 v^*)\right| = \left|\varphi(v p_0)\right| \le \|\varphi\|\|v p_0\| \le \|\varphi\|.\end{aligned}$$

Now taking supremum over finite subsets $J_0 \subset J$, we find that $x \in \mathsf{B}_1(\mathcal{H})$ and $\|x\|_1 \le \|\varphi\|$.

As $\mathsf{B}_2(\mathcal{H})$ is a dense subspace of $\mathsf{B}_0(\mathcal{H})$ (it contains $\mathcal{F}(\mathcal{H})$), we see that $\varphi = \varphi_x$. In particular $x \mapsto \varphi_x$ is a bijective isometry. □

The norm

$$\|x\|_1 = \operatorname{Tr}\left(|x|\right), \qquad x \in \mathsf{B}_1(\mathcal{H})$$

is called the *trace norm* on $\mathsf{B}_1(\mathcal{H})$.

The conclusion of Theorem 6.16 is a "non-commutative analog" of the well known equality $\ell_1 = {c_0}^*$. Similarly the "commutative" equality ${\ell_1}^* = \ell_\infty$ has a corresponding result on the level of operators on $\mathcal{H}$:

Theorem 6.17

For $y \in \mathsf{B}(\mathcal{H})$ let

$$\psi_y : \mathsf{B}_1(\mathcal{H}) \ni x \longmapsto \operatorname{Tr}(yx) \in \mathbb{C}.$$

Then $\psi_y \in \mathsf{B}_1(\mathcal{H})^*$ and $y \mapsto \psi_y$ is an isometry of $\mathsf{B}(\mathcal{H})$ onto $\mathsf{B}_1(\mathcal{H})^*$.

Proof

For $x \in \mathsf{B}_1(\mathcal{H})$ and $y \in \mathsf{B}(\mathcal{H})$ we have $\left|\operatorname{Tr}(yx)\right| \le \|y\|\|x\|_1$, so $\psi_y \in \mathsf{B}_1(\mathcal{H})^*$ and $\|\psi_y\| \le \|y\|$.

Let us take an arbitrary $\psi \in \mathsf{B}_1(\mathcal{H})^*$. Now define a sesquilinear form F on $\mathcal{H}$ putting

$$F(\eta, \xi) = \psi\big(|\xi\rangle\langle\eta|\big), \qquad \xi, \eta \in \mathcal{H}.$$

Then F is bounded[3]:

$$\left|F(\eta, \xi)\right| \le \|\psi\| \big\| |\xi\rangle\langle\eta| \big\|_1 = \|\psi\|\|\xi\|\|\eta\|, \qquad \xi, \eta \in \mathcal{H},$$

and hence there exists $y \in \mathsf{B}(\mathcal{H})$ such that

$$\langle\eta|y\xi\rangle = F(\eta, \xi), \qquad \xi, \eta \in \mathcal{H}.$$

Now let $x \in \mathsf{B}_1(\mathcal{H})$ be self-adjoint. Since x is compact (and self-adjoint), we can write x in the form of a norm-convergent series

$$x = \sum_{n=1}^{\infty} \mu_n |\psi_n\rangle\langle\psi_n| \tag{6.11}$$

for some orthonormal system $\{\psi_n\}_{n\in\mathbb{N}}$ and a sequence $(\mu_n)_{n\in\mathbb{N}}$ of real numbers. By considering an orthonormal basis obtained by completion of the system $\{\psi_n\}_{n\in\mathbb{N}}$ we easily see that $\sum_{n=1}^{\infty} |\mu_n| = \operatorname{Tr}\left(|x|\right) < +\infty$. It follows from this that the series (6.11) converges also for the trace norm. Therefore

$$\begin{aligned}\operatorname{Tr}(yx) &= \sum_{n=1}^{\infty} \langle\psi_n|yx\psi_n\rangle = \sum_{n=1}^{\infty} \mu_n \langle\psi_n|y\psi_n\rangle \\ &= \sum_{n=1}^{\infty} \mu_n F(\psi_n, \psi_n) = \sum_{n=1}^{\infty} \mu_n \psi\big(|\psi_n\rangle\langle\psi_n|\big) \\ &= \psi\left(\sum_{n=1}^{\infty} \mu_n |\psi_n\rangle\langle\psi_n|\right) = \psi(x)\end{aligned}$$

---

[3]For $\xi, \eta \in \mathcal{H}, \eta \neq 0$ we have

$$\big| |\xi\rangle\langle\eta| \big|^2 = \big(|\xi\rangle\langle\eta|\big)^* |\xi\rangle\langle\eta| = |\eta\rangle\langle\xi| \cdot |\xi\rangle\langle\eta| = \|\xi\|^2 |\eta\rangle\langle\eta| = \|\xi\|^2\|\eta\|^2 |\zeta\rangle\langle\zeta|,$$

where $\zeta = \frac{1}{\|\eta\|}\eta$. Thus $\big| |\xi\rangle\langle\eta| \big| = \|\xi\|\|\eta\| |\zeta\rangle\langle\zeta|$ and

$$\big\| \big| |\xi\rangle\langle\eta| \big| \big\|_1 = \operatorname{Tr}\big(\big| |\xi\rangle\langle\eta| \big|\big) = \|\xi\|\|\eta\| \operatorname{Tr}\big(|\zeta\rangle\langle\zeta|\big) = \|\xi\|\|\eta\|.$$

This means that $\psi(x) = \psi_y(x)$ for all self-adjoint $x \in \mathsf{B}_1(\mathcal{H})$, and consequently $\psi = \psi_y$. This way we have shown that

$$\mathsf{B}(\mathcal{H}) \ni y \longmapsto \psi_y \in \mathsf{B}_1(\mathcal{H})^*$$

is a contractive bijection.

To see that $\|\psi_y\| = \|y\|$ let us take $\varepsilon > 0$ and $\xi \in \mathcal{H}$ of norm 1 satisfying $\|y\xi\| > \|y\| - \varepsilon$. Furthermore let $\phi = \frac{1}{\|y\xi\|} y\xi$. Now let us complete the orthonormal system $\{\phi\}$ to a basis $\{\phi_j\}_{j\in J}$ of $\mathcal{H}$. Then, putting $x = |\xi\rangle\langle\phi|$, we have $\|x\|_1 = 1$ and, moreover,

$$\begin{aligned} |\psi_y(x)| = |\operatorname{Tr}(yx)| &= \Big|\sum_{j\in J} \langle\phi_j | yx\phi_j\rangle\Big| \\ &\geq |\langle\phi | yx\phi\rangle| = |\langle\phi | y\xi\rangle| \\ &= \tfrac{1}{\|y\xi\|}\langle y\xi | y\xi\rangle = \|y\xi\| > \|y\| - \varepsilon \end{aligned}$$

which means that $\|\psi_y\| \geq \|y\|$. □

6.3 Hilbert-Schmidt Operators on L_2

Let (Ω, μ) be a σ-finite measure space such that the space $L_2(\Omega, \mu)$ is separable and let $k \in L_2(\Omega \times \Omega, \mu \otimes \mu)$. By Fubini's theorem for almost all $\omega_1 \in \Omega$ the function $k(\omega_1, \cdot)$ is square-integrable and integrating with over the variable ω_1 the integral of the square of the absolute value of this function we obtain

$$\int_\Omega \Big(\int_\Omega |k(\omega_1, \omega_2)|^2 \, d\mu(\omega_2)\Big) d\mu(\omega_1) = \|k\|_2^2.$$

In particular, for any $\psi \in L_2(\Omega, \mu)$ the integral

$$\int_\Omega k(\omega_1, \omega_2)\psi(\omega_2)\, d\mu(\omega_2)$$

makes sense for almost all ω_1 and the resulting function of ω_1 satisfies

$$\begin{aligned} &\int_\Omega \Big|\int_\Omega k(\omega_1, \omega_2)\psi(\omega_2)\, d\mu(\omega_2)\Big|^2 d\mu(\omega_1) \\ &\qquad\qquad \leq \int_\Omega \int_\Omega |k(\omega_1, \omega_2)|^2 \, d\mu(\omega_2) \|\psi\|_2^2 \, d\mu(\omega_1) = \|k\|_2^2 \|\psi\|_2^2. \end{aligned}$$

From this we infer that the formula

$$(t_k\psi)(\omega_1) = \int_\Omega k(\omega_1, \omega_2)\psi(\omega_2)\, d\mu(\omega_2), \qquad \omega_1 \in \Omega$$

defines an element $t_k\psi$ of $L_2(\Omega, \mu)$ and the resulting map

$$t_k : L_2(\Omega, \mu) \ni \psi \longmapsto t_k\psi \in L_2(\Omega, \mu)$$

is linear and bounded with

$$\|t_k\| \leq \|k\|_2. \tag{6.12}$$

The operator t_k is called an *integral operator*, while the function k is the *integral kernel* of the operator t_k.

Theorem 6.18
The range of the map

$$L_2(\Omega \times \Omega, \mu \otimes \mu) \ni k \longmapsto t_k \in \mathsf{B}\big(L_2(\Omega, \mu)\big)$$

coincides with $\mathsf{B}_2\big(L_2(\Omega, \mu)\big)$ *and the resulting operator*

$$L_2(\Omega \times \Omega, \mu \otimes \mu) \longrightarrow \mathsf{B}_2\big(L_2(\Omega, \mu)\big)$$

is unitary.

Proof
Let $\{\varphi_i\}_{i\in I}$ be an orthonormal basis of $L_2(\Omega, \mu)$. Then the system $\{\varphi_i \otimes \overline{\varphi_j}\}_{i,j\in I}$ is an orthonormal basis of

$$L_2(\Omega, \mu) \otimes L_2(\Omega, \mu) \cong L_2(\Omega \times \Omega, \mu \otimes \mu)$$

(see Appendix A.3). This gives us a Fourier expansion

$$k = \sum_{i,j\in I} \alpha_{i,j}\varphi_i \otimes \overline{\varphi_j}.$$

Let $\mathcal{I}$ be the family of finite subsets of I. For $A \in \mathcal{I}$ let

$$k_A = \sum_{i,j\in A} \alpha_{i,j}\varphi_i \otimes \overline{\varphi_j}$$

and consider the operator t_{k_A}. For each $\psi \in L_2(\Omega, \mu)$ we have

$$(t_{k_A}\psi)(\omega_1) = \sum_{i,j\in A} \alpha_{i,j} \int_\Omega \overline{\varphi_j}(\omega_2)\psi(\omega_2)\, d\mu(\omega_2)\, \varphi_i(\omega_1),$$

which means that

$$t_{k_A} = \sum_{i,j\in A} \alpha_{i,j} |\varphi_i\rangle\langle\varphi_j|.$$

In particular $t_{k_A} \in \mathcal{F}\big(L_2(\Omega, \mu)\big) \subset \mathsf{B}_2\big(L_2(\Omega, \mu)\big)$. Furthermore, it follows from the estimate (6.12) that

$$\|t_k - t_{k_A}\| = \|t_{k-k_A}\| \leq \|k - k_A\|_2 \xrightarrow[A\in\mathcal{I}]{} 0,$$

so that t_k is a compact operator. In addition, for any $s \in I$

$$t_k\varphi_s = \lim_{A\in\mathcal{I}} t_{k_A}\varphi_s = \sum_{i\in I} \alpha_{i,s}\varphi_i.$$

Thus

$$\|t_k\varphi_s\|_2^2 = \sum_{i\in I} |\alpha_{i,s}|^2$$

and

$$\mathrm{Tr}(t_k{}^*t_k) = \sum_{s\in I} \|t_k\varphi_s\|_2^2 = \sum_{i,s\in I} |\alpha_{i,s}|^2 = \|k\|_2^2.$$

The above arguments show that $k \mapsto t_k$ is an isometry from $L_2(\Omega \times \Omega, \mu \otimes \mu)$ into $\mathsf{B}_2\big(L_2(\Omega, \mu)\big)$. Moreover, its range contains a dense subset $\mathcal{F}\big(L_2(\Omega, \mu)\big)$, and therefore this map must be unitary. □

Notes

Trace class and Hilbert-Schmidt operators are very useful objects both in pure mathematics and in theoretical physics, where the duality between $\mathsf{B}_1(\mathcal{H})$ and $\mathsf{B}(\mathcal{H})$ provides a description of so called *mixed states* of a quantum system. As for other applications, let us point out that Theorem 6.18 can be regarded as a criterion of compactness. More precisely, if t is a bounded operator on $L_2(\Omega, \mu)$ which can be written as an integral operator with square-integrable kernel then t belongs to the class of Hilbert-Schmidt operators and hence is compact. Examples and exercises related to the trace, trace class operators and Hilbert-Schmidt operators can be found in [Mau, Chapter VII], [Ped, Section 3.4], [ReSi$_1$, Chapter VI] and the problem book [Hal].

Functional Calculus for Families of Operators

P. Sołtan, *A Primer on Hilbert Space Operators*, Compact Textbooks in Mathematics,
https://doi.org/10.1007/978-3-319-92061-0_7

In this chapter we will extend functional calculus to families of commuting self-adjoint operators. As an application of this extension we will be able to introduce in ▶ Sect. 7.4 functional calculus for normal operators. This will be the only part of the book in which we will require some results of the theory of Banach algebras, or more specifically, C*-algebras. These have been gathered in Appendix A.5.

7.1 Holomorphic Functional Calculus

Let $a_1, \ldots, a_n \in \mathsf{B}(\mathcal{H})$ be pairwise commuting operators. For a function f holomorphic on a neighborhood of $\bigtimes_{i=1}^{n} \sigma(a_i)$ in $\mathbb{C}^n$ we define

$$f(a_1, \ldots, a_n)$$
$$= \left(\tfrac{1}{2\pi \mathrm{i}}\right)^n \oint_{\Gamma_n} \cdots \oint_{\Gamma_1} f(\lambda_1, \ldots, \lambda_n)(\lambda_1 \mathbb{1} - a_1)^{-1} \cdots (\lambda_n \mathbb{1} - a_n)^{-1} d\lambda_1 \ldots d\lambda_n,$$

where $\Gamma_1, \ldots, \Gamma_n$ are positively oriented curves (in $\mathbb{C}$) such that for each i the curve Γ_i surrounds $\sigma(a_i)$ and the product $\bigtimes_{i=1}^{n} \Gamma_i$ lies in the domain of holomorphy of f. Of course the value of the integral does not depend on the choice of the curves $\Gamma_1, \ldots, \Gamma_n$.

Just as in the case of a single operator, we will denote by $\mathscr{H}\left(\bigtimes_{i=1}^{n} \sigma(a_i)\right)$ the algebra of functions holomorphic on a neighborhood of $\bigtimes_{i=1}^{n} \sigma(a_i)$.

Theorem 7.1
The map

$$\mathscr{H}\left(\bigtimes_{i=1}^{n} \sigma(a_i)\right) \ni f \longmapsto f(a_1,\ldots,a_n) \in \mathsf{B}(\mathcal{H}) \tag{7.1}$$

is a unital homomorphism.

Proof
It is clear that the map (7.1) is linear. Now we check that for any $k \in \{1,\ldots,n\}$ and $N \in \mathbb{Z}_+$ if $f(\lambda_1,\ldots,\lambda_n) = \lambda_k^N$ then $f(a_1,\ldots,a_n) = a_k^N$: choose $\Gamma_1,\ldots,\Gamma_n$, so that for each i the curve Γ_i lies in

$$\{z \in \mathbb{C} \,|\, |z| > \|a_i\|\}.$$

We have

$$\begin{aligned}
&\left(\tfrac{1}{2\pi \mathrm{i}}\right)^n \oint_{\Gamma_n}\cdots\oint_{\Gamma_1} \lambda_k^N (\lambda_1 \mathbb{1} - a_1)^{-1}\cdots(\lambda_n\mathbb{1} - a_n)^{-1} d\lambda_1 \ldots d\lambda_n \\
&= \left(\tfrac{1}{2\pi \mathrm{i}}\right)^n \oint_{\Gamma_n}\cdots\oint_{\Gamma_1} \lambda_k^N \left(\sum_{i=0}^{\infty} \lambda_1^{-i-1} a_1^i\right)\cdots\left(\sum_{i=0}^{\infty} \lambda_n^{-i-1} a_n^i\right) d\lambda_1 \ldots d\lambda_n \\
&= \prod_{j\neq k}\left(\tfrac{1}{2\pi \mathrm{i}} \oint_{\Gamma_j}\sum_{i=0}^{\infty} \lambda_j^{-i-1} a_j^i \, d\lambda_j\right)\cdot\left(\tfrac{1}{2\pi \mathrm{i}} \oint_{\Gamma_k}\sum_{i=0}^{\infty} \lambda_j^{N-i-1} a_k^i \, d\lambda_k\right) \\
&= \prod_{j\neq k}\left(\sum_{i=0}^{\infty} \tfrac{1}{2\pi \mathrm{i}} \oint_{\Gamma_j} \lambda_j^{-i-1}\, d\lambda_j \, a_j^i\right)\cdot\left(\sum_{i=0}^{\infty} \tfrac{1}{2\pi \mathrm{i}} \oint_{\Gamma_k} \lambda_j^{N-i-1}\, d\lambda_k \, a_k^i\right) = a_k^N.
\end{aligned}$$

It remains to prove multiplicativity of the map (7.1). First we notice that resolvents of commuting operators commute. To see this note that since $a_1,\ldots,a_n$ pairwise commute, for any $\lambda_i \in \rho(a_i)$ the product of resolvents $(\lambda_i\mathbb{1} - a_i)^{-1}$ in arbitrary order is the inverse of $\prod_{i=1}^{n}(\lambda_i\mathbb{1} - a_i)$. The remaining calculation is similar to the one carried out in the proof of Theorem 4.12: let $f, g \in \mathscr{H}\left(\bigtimes_{i=1}^{n} \sigma(a_i)\right)$ and let $\Gamma_1,\ldots,\Gamma_n$ and $\Gamma'_1,\ldots,\Gamma'_n$ be positively oriented curves surrounding the spectra of $\sigma(a_1),\ldots,\sigma(a_n)$ such that the sets $\bigtimes_{i=1}^{n}\Gamma_i$ and $\bigtimes_{i=1}^{n}\Gamma'_i$ lie in the intersection of domains of holomorphy of f and g. Furthermore we ask that for each i the curve Γ'_i lie outside of Γ_i.

Using the resolvent identity (1.2) we compute

$$
\begin{aligned}
f(a_1,\dots,a_n)g(a_1,\dots,a_n)
&= \left(\tfrac{1}{2\pi \mathrm{i}}\right)^n \oint_{\Gamma_n}\cdots\oint_{\Gamma_1} f(\lambda_1,\dots,\lambda_n)\prod_{i=1}^{n}(\lambda_i\mathbb{1}-a_i)^{-1}d\lambda_1\dots d\lambda_n\\
&\quad\cdot\left(\tfrac{1}{2\pi \mathrm{i}}\right)^n \oint_{\Gamma_n'}\cdots\oint_{\Gamma_1'} g(\mu_1,\dots,\mu_n)\prod_{j=1}^{n}(\mu_j\mathbb{1}-a_j)^{-1}d\mu_1\dots d\mu_n\\
&= \left(\tfrac{1}{2\pi \mathrm{i}}\right)^{2n} \oint_{\Gamma_n}\cdots\oint_{\Gamma_1}\oint_{\Gamma_n'}\cdots\oint_{\Gamma_1'} f(\lambda_1,\dots,\lambda_n)g(\mu_1,\dots,\mu_n)\\
&\quad\prod_{i=1}^{n}\left((\lambda_i\mathbb{1}-a_i)^{-1}(\mu_i\mathbb{1}-a_i)^{-1}\right)d\lambda_1\dots d\lambda_n\,d\mu_1\dots d\mu_n\\
&= \left(\tfrac{1}{2\pi \mathrm{i}}\right)^{2n} \oint_{\Gamma_n}\cdots\oint_{\Gamma_1}\oint_{\Gamma_n'}\cdots\oint_{\Gamma_1'} f(\lambda_1,\dots,\lambda_n)g(\mu_1,\dots,\mu_n)\\
&\quad\prod_{i=1}^{n}\left(\tfrac{1}{\mu_i-\lambda_i}\left((\lambda_i\mathbb{1}-a_i)^{-1}-(\mu_i\mathbb{1}-a_i)^{-1}\right)\right)d\lambda_1\dots d\lambda_n\,d\mu_1\dots d\mu_n.
\end{aligned}
$$

The quantity

$$
\prod_{i=1}^{n}\left(\tfrac{1}{\mu_i-\lambda_i}\left((\lambda_i\mathbb{1}-a_i)^{-1}-(\mu_i\mathbb{1}-a_i)^{-1}\right)\right)
$$

is a linear combination of terms which are products of a number of factors $(\lambda_i\mathbb{1}-a_i)^{-1}$ and a number of factors of the form $(\mu_j\mathbb{1}-a_j)^{-1}$. For example if $n=3$ we have

$$
\begin{aligned}
&\prod_{i=1}^{3}\left(\tfrac{1}{\mu_i-\lambda_i}\left((\lambda_i\mathbb{1}-a_i)^{-1}-(\mu_i\mathbb{1}-a_i)^{-1}\right)\right)\\
&\qquad= \tfrac{1}{\mu_1-\lambda_1}\tfrac{1}{\mu_2-\lambda_2}\tfrac{1}{\mu_3-\lambda_3}\big((\lambda_1\mathbb{1}-a_1)^{-1}(\lambda_2\mathbb{1}-a_2)^{-1}(\lambda_3\mathbb{1}-a_3)^{-1}\\
&\qquad\qquad-(\mu_1\mathbb{1}-a_1)^{-1}(\lambda_2\mathbb{1}-a_2)^{-1}(\lambda_3\mathbb{1}-a_3)^{-1}\\
&\qquad\qquad-(\lambda_1\mathbb{1}-a_1)^{-1}(\mu_2\mathbb{1}-a_2)^{-1}(\lambda_3\mathbb{1}-a_3)^{-1}\\
&\qquad\qquad-(\lambda_1\mathbb{1}-a_1)^{-1}(\lambda_2\mathbb{1}-a_2)^{-1}(\mu_3\mathbb{1}-a_3)^{-1}\\
&\qquad\qquad+(\mu_1\mathbb{1}-a_1)^{-1}(\mu_2\mathbb{1}-a_2)^{-1}(\lambda_3\mathbb{1}-a_3)^{-1}\\
&\qquad\qquad+(\mu_1\mathbb{1}-a_1)^{-1}(\lambda_2\mathbb{1}-a_2)^{-1}(\mu_3\mathbb{1}-a_3)^{-1}\\
&\qquad\qquad+(\lambda_1\mathbb{1}-a_1)^{-1}(\mu_2\mathbb{1}-a_2)^{-1}(\mu_3\mathbb{1}-a_3)^{-1}\\
&\qquad\qquad-(\mu_1\mathbb{1}-a_1)^{-1}(\mu_2\mathbb{1}-a_2)^{-1}(\mu_3\mathbb{1}-a_3)^{-1}\big).
\end{aligned}
$$

The curves $\Gamma_1, \ldots, \Gamma_n$ and $\Gamma'_1, \ldots, \Gamma'_n$ are chosen so that the integral of any term containing at least one $(\mu_i \mathbb{1} - a_i)^{-1}$ is equal to 0. To see this, consider e.g. the term

$$\oint_{\Gamma_3}\oint_{\Gamma_2}\oint_{\Gamma_1}\oint_{\Gamma'_3}\oint_{\Gamma'_2}\oint_{\Gamma'_1} f(\lambda_1, \lambda_2, \lambda_3) g(\mu_1, \mu_2, \mu_3) \frac{1}{\mu_1-\lambda_1}\frac{1}{\mu_2-\lambda_2}\frac{1}{\mu_3-\lambda_3}$$
$$(\mu_1\mathbb{1} - a_1)^{-1}(\lambda_2\mathbb{1} - a_2)^{-1}(\lambda_3\mathbb{1} - a_3)^{-1} d\lambda_1 d\lambda_2 d\lambda_3 d\mu_1 d\mu_2 d\mu_3$$
$$= \oint_{\Gamma_3}\oint_{\Gamma_2}\oint_{\Gamma'_3}\oint_{\Gamma'_2}\oint_{\Gamma'_1}\left(\oint_{\Gamma_1}\frac{f(\lambda_1,\lambda_2,\lambda_3)}{\mu_1-\lambda_1}d\lambda_1\right)\frac{1}{\mu_2-\lambda_2}\frac{1}{\mu_3-\lambda_3} g(\mu_1, \mu_2, \mu_3)$$
$$(\mu_1\mathbb{1} - a_1)^{-1}(\lambda_2\mathbb{1} - a_2)^{-1}(\lambda_3\mathbb{1} - a_3)^{-1} d\lambda_2 d\lambda_3 d\mu_1 d\mu_2 d\mu_3.$$

The integral $\oint_{\Gamma_1} \frac{f(\lambda_1,\lambda_2,\lambda_3)}{\mu_1-\lambda_1} d\lambda_1$ equals 0 because the function

$$\lambda_1 \longmapsto \frac{f(\lambda_1,\lambda_2,\lambda_3)}{\mu_1-\lambda_1}$$

is holomorphic in a contractible region containing the curve Γ_1.

It follows that we can eliminate from the expression we established for $f(a_1, \ldots, a_n)$ $g(a_1, \ldots, a_n)$ all terms containing at least one $(\mu_i \mathbb{1} - a_i)^{-1}$. This way we obtain

$$f(a_1, \ldots, a_n) g(a_1, \ldots, a_n)$$
$$= \left(\frac{1}{2\pi \mathrm{i}}\right)^{2n} \oint_{\Gamma_n}\cdots\oint_{\Gamma_1}\oint_{\Gamma'_n}\cdots\oint_{\Gamma'_1} f(\lambda_1, \ldots, \lambda_n) g(\mu_1, \ldots, \mu_n)$$
$$\prod_{i=1}^{n}\frac{1}{\mu_i-\lambda_i}\prod_{j=1}^{n}(\lambda_j\mathbb{1} - a_j)^{-1} d\lambda_1 \ldots d\lambda_n d\mu_1 \ldots d\mu_n$$
$$= \left(\frac{1}{2\pi \mathrm{i}}\right)^{n} \oint_{\Gamma_n}\cdots\oint_{\Gamma_1} f(\lambda_1, \ldots, \lambda_n)\left(\left(\frac{1}{2\pi \mathrm{i}}\right)^{n}\oint_{\Gamma'_n}\cdots\oint_{\Gamma'_1}\frac{g(\mu_1,\ldots,\mu_n)}{\prod_{i=1}^{n}(\mu_i-\lambda_i)}d\mu_1 \ldots d\mu_n\right)$$
$$\prod_{j=1}^{n}(\lambda_j\mathbb{1} - a_j)^{-1} d\lambda_1 \ldots d\lambda_n$$
$$= \left(\frac{1}{2\pi \mathrm{i}}\right)^{n} \oint_{\Gamma_n}\cdots\oint_{\Gamma_1} f(\lambda_1, \ldots, \lambda_n) g(\lambda_1, \ldots, \lambda_n)\prod_{j=1}^{n}(\lambda_j\mathbb{1} - a_j)^{-1} d\lambda_1 \ldots d\lambda_n$$
$$= (fg)(a_1, \ldots, a_n)$$

by the multidimensional version of Cauchy's formula. □

The holomorphic functional calculus for finite families of commuting self-adjoint operators we defined above has several properties analogous to those of functional calculus for one operator. The next lemma describes a weak version of the spectral mapping theorem:

Lemma 7.2 *Let* $a_1, \dots, a_n \in \mathsf{B}(\mathcal{H})$ *be pairwise commuting operators and let* P *be a polynomial in* n *variables. Then*

$$\begin{aligned}\sigma\big(P(a_1,\dots,a_n)\big) &\subset P\Big(\bigtimes_{i=1}^{n}\sigma(a_i)\Big)\\ &= \big\{P(\lambda_1,\dots,\lambda_n)\,\big|\,\lambda_i\in\sigma(a_i),\ i=1,\dots,n\big\}.\end{aligned}$$

Proof
Take $\mu \notin P\Big(\bigtimes_{i=1}^{n}\sigma(a_i)\Big)$. Then the function

$$f(z_1,\dots,z_n) = \big(\mu - P(z_1,\dots,z_n)\big)^{-1}$$

is holomorphic on a neighborhood of $\bigtimes_{i=1}^{n}\sigma(a_i)$ in $\mathbb{C}^n$ and we have

$$\big(\mu\mathbb{1} - P(a_1,\dots,a_n)\big)f(a_1,\dots,a_n) = f(a_1,\dots,a_n)\big(\mu\mathbb{1} - P(a_1,\dots,a_n)\big) = \mathbb{1}.$$

Consequently $\mu \notin \sigma\big(P(a_1,\dots,a_n)\big)$. □

Lemma 7.2 will be of crucial importance in the next section where we define continuous functional calculus for families of commuting self-adjoint operators.

7.2 Continuous Functional Calculus

Theorem 7.3
Let $x_1,\dots,x_n \in \mathsf{B}(\mathcal{H})$ *be pairwise commuting self-adjoint operators. Then there exists a unique unital* $*$*-homomorphism*

$$\mathsf{C}\Big(\bigtimes_{i=1}^{n}\sigma(a_i)\Big) \longrightarrow \mathsf{B}(\mathcal{H}),$$

denoted by $f \mapsto f(x_1,\dots,x_n)$*, such that if* $\mathrm{pr}_j : \bigtimes_{i=1}^{n}\sigma(a_i) \to \sigma(a_j)$ *is the projection onto* j*th coordinate then*

$$\mathrm{pr}_j(x_1,\dots,x_n) = x_j, \qquad j = 1,\dots,n.$$

Proof

If a $*$-homomorphism described in the theorem exists then it must assign the value

$$P(x_1, \dots, x_n) = \sum_{i_1,\dots,i_n} \alpha_{i_1,\dots,i_n} x_1^{i_1} \cdots x_n^{i_n} \tag{7.2}$$

to the polynomial

$$P(\lambda_1, \dots, \lambda_n) = \sum_{i_1,\dots,i_n} \alpha_{i_1,\dots,i_n} \lambda_1^{i_1} \cdots \lambda_n^{i_n}.$$

Let us, therefore, take (7.2) as the definition of $P(x_1, \dots, x_n)$. It is easy to see that the right hand side of (7.2) is a normal operator, and so, by Lemma 7.2

$$\begin{aligned} \|P(x_1, \dots, x_n)\| &= \left|\sigma\left(P(x_1, \dots, x_n)\right)\right| \\ &\le \sup\left\{|P(\lambda_1, \dots, \lambda_n)| \,\middle|\, \lambda_i \in \sigma(x_i),\ i = 1, \dots, n\right\} = \|P\|_\infty, \end{aligned} \tag{7.3}$$

where $\|\cdot\|_\infty$ denotes the uniform norm on $\mathrm{C}\left(\bigtimes_{i=1}^{n} \sigma(a_i)\right)$. In particular $P(x_1, \dots, x_n)$ depends only on the values of P on the set $\bigtimes_{i=1}^{n} \sigma(a_i)$.

It is immediate that the map

$$P \longmapsto P(x_1, \dots, x_n)$$

is a $*$-homomorphism and the estimate (7.3) shows that it extends uniquely to a $*$-homomorphism $\mathrm{C}\left(\bigtimes_{i=1}^{n} \sigma(a_i)\right) \to \mathsf{B}(\mathcal{H})$ satisfying the conditions of the theorem. □

7.3 Joint Spectrum

Consider a family $x_1, \dots, x_n \in \mathsf{B}(\mathcal{H})$ of pairwise commuting self-adjoint operators and the compact space $X = \bigtimes_{i=1}^{n} \sigma(x_i)$. Let

$$\mathsf{J} = \left\{f \in \mathrm{C}(X) \,\middle|\, f(x_1, \dots, x_n) = 0\right\}.$$

Clearly J is an ideal in $\mathrm{C}(X)$. It is shown in Appendix A.5.2 that we can associate with J a closed subset $Y \subset X$ such that

$$\mathsf{J} = \mathrm{C}_0(X \setminus Y) \quad \text{and} \quad \mathrm{C}(X)/\mathsf{J} \cong \mathrm{C}(Y).$$

The mapping $\mathrm{C}(X) \ni f \mapsto f(x_1, \dots, x_n)$ factorizes through $\mathrm{C}(X)/\mathsf{J}$ (since J is, by definition, its kernel), and so we obtain an isometric unital $*$-homomorphism

$$\mathrm{C}(Y) \longrightarrow \mathsf{B}(\mathcal{H})$$

which we call the *continuous functional calculus* for the operators $x_1, \ldots, x_n$. The set Y is the *joint spectrum* of the operators $x_1, \ldots, x_n$ which we denote by the symbol $\sigma(x_1, \ldots, x_n)$.

Clearly, a point $(\mu_1, \ldots, \mu_n) \in \bigtimes_{i=1}^{n} \sigma(x_i)$ lies outside of $\sigma(x_1, \ldots, x_n)$ if and only if there exists a function $f \in \mathrm{C}\left(\bigtimes_{i=1}^{n} \sigma(x_i)\right)$ such that

$$f(\mu_1, \ldots, \mu_n) \neq 0 \quad \text{and} \quad f(x_1, \ldots, x_n) = 0.$$

7.4 Functional Calculus for Normal Operators

Let $x \in \mathrm{B}(\mathcal{H})$. Define $\operatorname{Re} x = \frac{1}{2}(x + x^*)$ and $\operatorname{Im} x = \frac{1}{2\mathrm{i}}(x - x^*)$. Then $\operatorname{Re} x$ and $\operatorname{Im} x$ are self-adjoint and

$$x = \operatorname{Re} x + \mathrm{i} \operatorname{Im} x. \tag{7.4}$$

The operators $\operatorname{Re} x$ and $\operatorname{Im} x$ are usually called the *real part* and *imaginary part* of x. Moreover, (7.4) is clearly the unique way to write x as a sum of a self-adjoint operator and an anti-self-adjoint operator (i.e. a self-adjoint operator multiplied by i). It is easy to check that x is normal if and only if its real and imaginary parts commute.

Proposition 7.4 *Let $x \in \mathrm{B}(\mathcal{H})$ be a normal operator. Then*

$$\sigma(x) = \big\{a + \mathrm{i}b \,\big|\, (a, b) \in \sigma(\operatorname{Re} x, \operatorname{Im} x)\big\}. \tag{7.5}$$

Proof

Take $(c, d) \notin \sigma(\operatorname{Re} x, \operatorname{Im} x)$. Then the function

$$f : (a, b) \longmapsto \tfrac{1}{(c+\mathrm{i}d)-(a+\mathrm{i}b)}$$

is continuous on $\sigma(\operatorname{Re} x, \operatorname{Im} x)$ and it easily follows that the operator

$$(c + \mathrm{i}d)\mathbb{1} - x = (c + \mathrm{i}d)\mathbb{1} - (\operatorname{Re} x + \mathrm{i} \operatorname{Im} x)$$

is invertible with inverse $f(\operatorname{Re} x, \operatorname{Im} x)$. This proves the containment "$\subset$" in (7.5).

Let $\lambda \in \rho(x)$ and put $r = \operatorname{Re} \lambda$, $s = \operatorname{Im} \lambda$. Suppose further that $(r, s) \in \sigma(\operatorname{Re} x, \operatorname{Im} x)$. For $\varepsilon > 0$ let f be a continuous function on $\sigma(\operatorname{Re} x, \operatorname{Im} x)$ such that $\|f\|_\infty \leq 1$, $f(r, s) = 1$ and the support of f is contained in the set

$$\big\{(p, g) \,\big|\, (p - r)^2 + (q - s)^2 \leq \varepsilon^2\big\}.$$

We have

$$f(\operatorname{Re} x, \operatorname{Im} x) = f(\operatorname{Re} x, \operatorname{Im} x)\big((r+\mathrm{i}s)\mathbb{1} - x\big)\big((r+\mathrm{i}s)\mathbb{1} - x\big)^{-1},$$

and hence

$$\big\|f(\operatorname{Re} x, \operatorname{Im} x)\big\| \leq \big\|f(\operatorname{Re} x, \operatorname{Im} x)\big((r+\mathrm{i}s)\mathbb{1} - x\big)\big\|\big\|\big((r+\mathrm{i}s)\mathbb{1} - x\big)^{-1}\big\|.$$

Note that

$$(r+\mathrm{i}s)\mathbb{1} - x = (r+\mathrm{i}s)\mathbb{1} - (\operatorname{Re} x + \mathrm{i}\operatorname{Im} x) = g(\operatorname{Re} x, \operatorname{Im} x),$$

where $g(p,q) = (r+\mathrm{i}s) - (p+\mathrm{i}q)$. Therefore

$$\begin{aligned}\big\|f(\operatorname{Re} x, \operatorname{Im} x)\big((r+\mathrm{i}s)\mathbb{1} - x\big)\big\| &= \big\|f(\operatorname{Re} x, \operatorname{Im} x)g(\operatorname{Re} x, \operatorname{Im} x)\big\| \\ &= \big\|(fg)(\operatorname{Re} x, \operatorname{Im} x)\big\| \leq \|fg\|_\infty \leq \varepsilon\end{aligned}$$

(as $|g| \leq \varepsilon$ on the support of f) and it follows that

$$\big\|f(\operatorname{Re} x, \operatorname{Im} x)\big\| \leq \varepsilon\big\|\big((r+\mathrm{i}s)\mathbb{1} - x\big)^{-1}\big\|. \tag{7.6}$$

But the right hand side (7.6) is arbitrarily small (as we vary ε), while $\big\|f(\operatorname{Re} x, \operatorname{Im} x)\big\| \geq 1$, because $f(r,s) = 1$. This contradiction shows that if $\lambda = r + \mathrm{i}s \notin \sigma(x)$ then $(r,s) \notin \sigma(\operatorname{Re} x, \operatorname{Im} x)$ and we get the containment "$\supset$" in (7.5). □

Given a normal operator $x \in \mathsf{B}(\mathcal{H})$ and a continuous function f on $\sigma(x)$ we can define

$$f(x) = \check{f}(\operatorname{Re} x, \operatorname{Im} x),$$

where $\check{f}(u,v) = f(u + \mathrm{i}v)$.

Corollary 7.5 *Let $x \in \mathsf{B}(\mathcal{H})$ be a normal operator. Then there exists a unique unital $*$-homomorphism $\mathsf{C}\big(\sigma(x)\big) \to \mathsf{B}(\mathcal{H})$ denoted by*

$$\mathsf{C}\big(\sigma(x)\big) \ni f \longmapsto f(x) \in \mathsf{B}(\mathcal{H})$$

such that if $f(\lambda) = \lambda$ for all $\lambda \in \sigma(x)$ then $f(x) = x$. Moreover $f \mapsto f(x)$ is an isometric $*$-isomorphism of $\mathsf{C}\big(\sigma(x)\big)$ onto $\mathrm{C}^*(x, \mathbb{1})$.

Proof

By uniqueness of the decomposition of x into its real and imaginary parts, the condition $f(x) = x$ for the identity function $f(\lambda) = \lambda$ is equivalent to the condition that

$$f_1(x) = \operatorname{Re} x \quad \text{and} \quad f_2(x) = \operatorname{Im} x,$$

where $f_1 = \operatorname{Re} f$ and $f_2 = \operatorname{Im} f$. The continuous functional calculus for the normal operator x provides an isometric $*$-homomorphism

$$\mathrm{C}\big(\sigma(\operatorname{Re} x, \operatorname{Im} x)\big) \ni f \longmapsto f(\operatorname{Re} x, \operatorname{Im} x) \in \mathrm{B}(\mathcal{H}) \tag{7.7}$$

satisfying this condition. Moreover, since polynomials in $\operatorname{Re} x$ and $\operatorname{Im} x$ span a dense subalgebra of $\mathrm{C}\big(\sigma(\operatorname{Re} x, \operatorname{Im} x)\big) = \mathrm{C}\big(\sigma(x)\big)$, the condition determines the homomorphism uniquely. Finally note that the range of (7.7) contains all polynomials in x and x^*. These are dense in $\mathrm{C}^*(x, \mathbb{1})$ and the range of a $*$-homomorphism between C^*-algebras is always closed by Theorem A.12, so it follows that the range of (7.7) is $\mathrm{C}^*(x, \mathbb{1})$. □

The $*$-isomorphism described in Corollary 7.5 is called the *continuous functional calculus* for the normal operator x. In the same way as for self-adjoint operators we prove the following properties of the continuous functional calculus:

- the spectral mapping theorem: $\sigma\big(f(x)\big) = f\big(\sigma(x)\big)$ for any normal $x \in \mathrm{B}(\mathcal{H})$ and $f \in \mathrm{C}\big(\sigma(x)\big)$,
- for any $g \in \mathrm{C}\big(\sigma(x)\big)$ the operator $g(x)$ is normal and for $f \in \mathrm{C}\big(\sigma\big(g(x)\big)\big)$ we have

$$f\big(g(x)\big) = (f \circ g)(x).$$

Using the isomorphism $\mathrm{C}\big(\sigma(x)\big) \ni f \mapsto f(x) \in \mathrm{C}^*(x, \mathbb{1})$ we can immediately prove an analog of Theorem 4.7 for normal operators:

Theorem 7.6
Let $x \in \mathrm{B}(\mathcal{H})$ be a normal operator. Then there exist a semifinite measure space (Ω, μ), an essentially bounded measurable function F on Ω and a unitary operator $u \in \mathrm{B}\big(L_2(\Omega, \mu), \mathcal{H}\big)$ such that

$$x = uM_F u^*.$$

The proof of Theorem 7.6 is identical to the proof of the analogous Theorem 4.7. The key element is the possibility of applying functions continuous on $\sigma(x)$ to x.

Theorem 7.6 allows us to extend the Borel functional calculus to the class of normal operators:

Theorem 7.7
Let $x \in \mathrm{B}(\mathcal{H})$ be a normal operator and denote by $\mathscr{B}\big(\sigma(x)\big)$ the algebra of bounded Borel functions on $\sigma(x)$. Then there exists a unique unital $*$-homomorphism $\mathscr{B} \to \mathrm{B}(\mathcal{H})$ denoted by

$$\mathscr{B}\big(\sigma(x)\big) \ni f \longmapsto f(x) \in \mathrm{B}(\mathcal{H})$$

(Continued)

Theorem 7.7 (continued)
such that
- *if* $f(\lambda) = \lambda$ *for all* $\lambda \in \sigma(x)$ *then* $f(x) = x$,
- *if* $(f_n)_{n\in\mathbb{N}}$ *is a uniformly bounded sequence of Borel functions converging pointwise to* f *then* $f_n(x) \xrightarrow[n\to\infty]{} f(x)$ *in strong topology.*

Moreover the above homomorphism extends the isomorphism $\mathrm{C}(\sigma(x)) \to \mathrm{C}^*(x, \mathbb{1})$ *given by the continuous functional calculus.*

Just as in the case of Theorem 7.6, the proof of Theorem 7.7 is merely a repetition of the steps taken to prove the analogous statement for self-adjoint operators (Theorem 4.10).

Let $x \in \mathrm{B}(\mathcal{H})$ be a normal operator. The extension of functional calculus for normal operators to all bounded Borel functions on $\sigma(x)$ makes it possible to define the spectral measure E_x associated to x by the formula

$$E_x(\Delta) = \chi_\Delta(x), \qquad \Delta \in \mathfrak{M},$$

where $\mathfrak{M}$ is the σ-algebra of Borel subsets of $\sigma(x)$. It is not hard to show (again, repeating the proof for self-adjoint operators, cf. ▶ Sect. 4.3) that for $f \in \mathscr{B}(\sigma(x))$ we have

$$f(x) = \int_{\sigma(x)} f(\lambda)\, dE_x(\lambda)$$

and, in particular,

$$x = \int_{\sigma(x)} \lambda\, dE_x(\lambda). \tag{7.8}$$

Finally, as is the case of self-adjoint operators, the spectral measure E_x such that (7.8) holds is unique.

Notes

Functional calculus and other versions of the spectral theorem for normal operators are usually introduced within the framework of Banach algebras and, in particular, Gelfand's theory of commutative Banach algebras ([Arv$_2$, Chapters 1 and 2], [Mau, Chapter VIII], [Ped, Chapter 4], [Rud$_2$, Chapters 10 and 11], [Zel]). Particularly relevant is the theory of commutative C*-algebras with the famous theorem of Gelfand and Naimark which says that any commutative C*-algebra with unit is isometrically $*$-isomorphic to an algebra of continuous functions on a uniquely determined compact space.

Our approach does not fully avoid Banach algebras, but we minimize their use favoring a more direct analysis of normal operators. Further developments include functional calculus for families of commuting normal operators which is developed by methods introduced in this chapter.

Part II
Unbounded Operators

Operators and Their Graphs

P. Sołtan, *A Primer on Hilbert Space Operators*, Compact Textbooks in Mathematics,
https://doi.org/10.1007/978-3-319-92061-0_8

Applications of operator theory in other branches of mathematics and in mathematical physics very often involve operators which are not bounded. This poses numerous difficulties whose source is, for the most part, the lack of useful algebraic structure on the set of unbounded operators. Our presentation of the theory of unbounded operators on a Hilbert space will focus on a few select issues and our preferred strategy for dealing with them will be to reduce them to questions about bounded operators. We will begin with some introductory information gathered in ▶ Sect. 8.1. In the following chapters we will introduce our key tool which we call the *z-transform* and later use this tool to extend various versions of the spectral theorem to unbounded self-adjoint operators. The final chapters will be devoted to several classical topics like self-adjoint extensions of symmetric operators and elements of the theory of one-parameter groups of unitary operators.

8.1 Basics of Unbounded Operators

Let $\mathcal{H}$ be a Hilbert space. The space $\mathcal{H} \oplus \mathcal{H}$ has a natural Hilbert space structure with scalar product:

$$\left\langle \begin{bmatrix} \xi \\ \eta \end{bmatrix} \middle| \begin{bmatrix} \psi \\ \phi \end{bmatrix} \right\rangle = \langle \xi | \psi \rangle + \langle \eta | \phi \rangle , \qquad \xi, \psi, \eta, \phi \in \mathcal{H}.$$

In this part of the book we will look at operators on $\mathcal{H}$ from a slightly different perspective than we have done so far. From now on a linear operator T on $\mathcal{H}$ will be defined on a vector subspace $\mathsf{D}(T) \subset \mathcal{H}$ called the *domain* of the operator T. In other words T is a linear map

$$T : \mathsf{D}(T) \longrightarrow \mathcal{H}.$$

We will not be assuming that $\mathsf{D}(T) = \mathcal{H}$, but except for some very general considerations and the material of ▶ Sect. 11.2, we will be assuming that $\mathsf{D}(T)$ is dense in $\mathcal{H}$. In the latter case we say that T is *densely defined*.

The *graph* of the operator T is the subspace

$$\mathsf{G}(T) = \left\{ \begin{bmatrix} \psi \\ T\psi \end{bmatrix} \middle| \, \psi \in \mathsf{D}(T) \right\} \subset \mathcal{H} \oplus \mathcal{H}.$$

We call T *closed* if the subspace $\mathsf{G}(T)$ is closed in $\mathcal{H} \oplus \mathcal{H}$. Note that for any $x \in \mathsf{B}(\mathcal{H})$ we have $\mathsf{D}(x) = \mathcal{H}$ and x is closed. Moreover, the closed graph theorem (Corollary A.7) says that if T is closed and $\mathsf{D}(T) = \mathcal{H}$ then T is bounded.

The condition that an operator T be closed replaces to a certain extent the condition of continuity. More precisely, it is easy to see that T is closed if and only if for any sequence $(\psi_n)_{n\in\mathbb{N}}$ of elements of $\mathsf{D}(T)$ we have

$$\Big(\lim_{n\to\infty} \psi_n = \psi, \ \lim_{n\to\infty} T\psi_n = \phi \Big) \Longrightarrow \Big(\psi \in \mathsf{D}(T), \ T\psi = \phi \Big).$$

It is worthwhile to characterize subspaces which are graphs of operators:

Proposition 8.1 *A subspace* $\mathsf{G} \subset \mathcal{H} \oplus \mathcal{H}$ *is a graph of a linear operator if and only if it does not contain non-zero vectors of the form*

$$\begin{bmatrix} 0 \\ \eta \end{bmatrix}. \tag{8.1}$$

Proof

Clearly a graph of a linear map cannot contain such vectors. On the other hand, if G does not contain non-zero vectors of the form (8.1) then the condition

$$\begin{bmatrix} \xi \\ \eta_1 \end{bmatrix}, \begin{bmatrix} \xi \\ \eta_2 \end{bmatrix} \in \mathsf{G}$$

implies $\eta_1 = \eta_2$, because G is a vector subspace. Therefore G is the graph of a map T defined on

$$\mathsf{D}(T) = \left\{ \xi \in \mathcal{H} \,\middle|\, \text{there exists } \eta \in \mathcal{H} \text{ such that } \begin{bmatrix} \xi \\ \eta \end{bmatrix} \in \mathsf{G} \right\}$$

by $T\xi = \eta$, where η is the unique vector such that $\left[\begin{smallmatrix} \xi \\ \eta \end{smallmatrix}\right] \in \mathsf{G}$. It is easy to check that this map is linear.

□

Vectors of the form (8.1) will be called *vertical*.

Let T be an operator on $\mathcal{H}$. We say that T is *closable* if $\overline{\mathsf{G}(T)}$ is a graph of an operator, i.e. when $\overline{\mathsf{G}(T)}$ does not contain non-zero vertical vectors. In this case the operator whose graph is $\overline{\mathsf{G}(T)}$ is called the *closure* of T and is denoted by $\overline{T}$.

Now if S and T are operators on $\mathcal{H}$ then S is an *extension* of T, or S *contains* T, if $\mathsf{G}(T) \subset \mathsf{G}(S)$. This means that $\mathsf{D}(T) \subset \mathsf{D}(S)$ and for $\psi \in \mathsf{D}(T)$ we have $S\psi = T\psi$. In this case we write $T \subset S$. In particular, if T is closable then $\overline{T}$ is an extension of T: $T \subset \overline{T}$.

Proposition 8.2 *An operator T is closable if and only if for any sequence $(\psi_n)_{n\in\mathbb{N}}$ of elements of $\mathsf{D}(T)$ we have*

$$\left(\lim_{n\to\infty} \psi_n = 0,\ \lim_{n\to\infty} T\psi_n = \phi\right) \Longrightarrow \left(\phi = 0\right). \tag{8.2}$$

Proof

If T is closable then condition (8.2) follows immediately from the fact that $\overline{T}$ is closed. On the other hand, if $\overline{\mathsf{G}(T)}$ contains a non-zero vertical vector $\left[\begin{smallmatrix}0\\ \eta\end{smallmatrix}\right] \in \mathsf{G}$ then there exists a sequence $(\psi_n)_{n\in\mathbb{N}}$ of elements of $\mathsf{D}(T)$ such that

$$\lim_{n\to\infty} \psi_n = 0 \quad \text{and} \quad \lim_{n\to\infty} T\psi_n = \eta \neq 0.$$

□

Proposition 8.3 *Let T be an operator on $\mathcal{H}$. Then T is densely defined if and only if $\mathsf{G}(T)^{\perp}$ does not contain any non-zero vectors of the form*

$$\begin{bmatrix}\xi\\ 0\end{bmatrix}. \tag{8.3}$$

Proof

Suppose a non-zero vector ξ is orthogonal to $\mathsf{D}(T)$. Then clearly the vector $\left[\begin{smallmatrix}\xi\\ 0\end{smallmatrix}\right]$ is orthogonal to $\mathsf{G}(T)$. Conversely, the condition

$$\begin{bmatrix}\xi\\ 0\end{bmatrix} \perp \mathsf{G}(T)$$

implies that ξ is orthogonal to $\mathsf{D}(T)$, so if $\xi \neq 0$ then $\mathsf{D}(T)$ is not dense in $\mathcal{H}$. □

Vectors of the form (8.3) will be called *horizontal*. Propositions 8.1 and 8.3 yield the following corollary:

Corollary 8.4 *A subspace $\mathsf{G} \subset \mathcal{H} \oplus \mathcal{H}$ is the graph of a closed densely defined operator if and only if G is closed, does not contain non-zero vertical vectors and $\mathsf{G}^{\perp}$ does not contain non-zero horizontal vectors.*

Let T be a closed operator. The graph $\mathsf{G}(T)$ of T is then a closed subspace of $\mathcal{H}\oplus\mathcal{H}$, so it is itself a Hilbert space. Moreover, the map

$$\mathsf{D}(T) \ni \psi \longmapsto \begin{bmatrix} \psi \\ T\psi \end{bmatrix} \in \mathsf{G}(T)$$

is bijective. We can therefore transfer the Hilbert space structure of $\mathsf{G}(T)$ onto $\mathsf{D}(T)$ defining the scalar product by

$$\langle \psi | \phi \rangle_T = \left\langle \begin{bmatrix} \psi \\ T\psi \end{bmatrix} \middle| \begin{bmatrix} \phi \\ T\phi \end{bmatrix} \right\rangle, \qquad \psi, \phi \in \mathsf{D}(T).$$

The resulting norm on $\mathsf{D}(T)$ is then given by

$$\|\psi\|_T = \sqrt{\|\psi\|^2 + \|T\psi\|^2}, \qquad \psi \in \mathsf{D}(T)$$

and it is called the *graph norm*. With this norm $\mathsf{D}(T)$ is a Hilbert space.

Remark 8.5 The graph norm can be defined for any linear operator—not necessarily closed one. In fact, it is easy to see that an operator T on $\mathcal{H}$ is closed if and only if $\mathsf{D}(T)$ is complete in the graph norm.

As an application of Remark 8.5 we can consider a generalization of the notion of a multiplication operator introduced in ▶ Sect. 4.1. Let (Ω, μ) be a semifinite measure space and let f be a measurable function on Ω. Define M_f to be the operator on $L_2(\Omega, \mu)$ such that

$$\mathsf{D}(M_f) = \{\psi \in L_2(\Omega, \mu) \mid f\psi \in L_2(\Omega, \mu)\}$$

and for $\psi \in \mathsf{D}(M_f)$ we have $M_f\psi = f\psi$.

Proposition 8.6 *M_f is a closed operator. If f is finite almost everywhere then M_f is densely defined.*

Proof

The graph norm on $\mathsf{D}(M_f)$ is given by

$$\|\psi\|_{M_f} = \Big(\int_\Omega |\psi|^2\, d\mu + \int_\Omega |f|^2|\psi|^2\, d\mu\Big)^{\frac{1}{2}} = \Big(\int_\Omega (1+|f|^2)|\psi|^2\, d\mu\Big)^{\frac{1}{2}},$$

i.e. it coincides with the norm of the Hilbert space $L_2(\Omega, (1+|f|^2)\mu)$. Clearly $\psi \in \mathsf{D}(M_f)$ if and only if $\psi \in L_2(\Omega, (1+|f|^2)\mu)$, so $\mathsf{D}(M_f)$ is a Hilbert space in its graph norm.

Now let χ_n be the characteristic function of $\{\omega \in \Omega \mid |f(\omega)| \leq n\}$. If f is finite almost everywhere then the sequence of functions $(\chi_n)_{n\in\mathbb{N}}$ converges pointwise to 1 almost everywhere. Moreover the range of each operator M_{χ_n} is contained in $\mathsf{D}(M_f)$ and by the dominated convergence theorem

$$\|\phi - M_{\chi_n}\phi\|_2 \xrightarrow[n\to\infty]{} 0, \qquad \phi \in L_2(\Omega, \mu).$$

It follows that $\mathsf{D}(M_f)$ is dense in $L_2(\Omega, \mu)$. □

8.2 Adjoint Operator

Let T be a densely defined operator on $\mathcal{H}$. Let $\mathsf{G} \subset \mathcal{H} \oplus \mathcal{H}$ be a subspace defined as follows:

$$\left(\begin{bmatrix} \xi \\ \eta \end{bmatrix} \in \mathsf{G} \right) \Longleftrightarrow \left(\forall\, \psi \in \mathsf{D}(T)\ \langle \xi | T\psi \rangle = \langle \eta | \psi \rangle \right). \tag{8.4}$$

Then G is a graph of an operator on $\mathcal{H}$. Indeed: if $\left[\begin{smallmatrix} 0 \\ \eta \end{smallmatrix}\right] \in \mathsf{G}$ then η is orthogonal to $\mathsf{D}(T)$, and so $\eta = 0$.

The operator whose graph is G defined by (8.4) is denoted by T^* and we call it the *adjoint* of T. Note that the description of continuous linear functionals on Hilbert spaces shows that the domain of T^* consists precisely of those vectors ξ for which the functional

$$\mathsf{D}(T) \ni \psi \longmapsto \langle \xi | T\psi \rangle$$

is bounded.

We say that an operator T is *self-adjoint* if $T = T^*$. This equality means, in particular, that $\mathsf{D}(T) = \mathsf{D}(T^*)$. When $T \subset T^*$ then T is called *symmetric* or *hermitian*.

Let us examine the graph of T^*. We have

$$\left(\begin{bmatrix} \xi \\ \eta \end{bmatrix} \in \mathsf{G}(T^*) \right) \Longleftrightarrow \left(\forall\, \psi \in \mathsf{D}(T)\ \left\langle \begin{bmatrix} \xi \\ \eta \end{bmatrix} \middle| \begin{bmatrix} T\psi \\ -\psi \end{bmatrix} \right\rangle = 0 \right).$$

In other words

$$\left(\begin{bmatrix} \xi \\ \eta \end{bmatrix} \in \mathsf{G}(T^*) \right) \Longleftrightarrow \left(\begin{bmatrix} \xi \\ \eta \end{bmatrix} \perp \begin{bmatrix} 0 & \mathbb{1} \\ -\mathbb{1} & 0 \end{bmatrix} \mathsf{G}(T) \right),$$

i.e.

$$\mathsf{G}(T^*) = \left(\begin{bmatrix} 0 & \mathbb{1} \\ -\mathbb{1} & 0 \end{bmatrix} \mathsf{G}(T) \right)^{\perp} = \begin{bmatrix} 0 & \mathbb{1} \\ -\mathbb{1} & 0 \end{bmatrix} (\mathsf{G}(T)^{\perp}) \tag{8.5}$$

(the last equality follows from the fact that $\begin{bmatrix} 0 & \mathbb{1} \\ -\mathbb{1} & 0 \end{bmatrix}$ is a unitary operator on $\mathcal{H} \oplus \mathcal{H}$). Since the orthogonal complement of any subset is closed, it follows that T^* is always closed. In particular a self-adjoint operator is automatically closed.

Proposition 8.7 *A densely defined operator T is closable if and only if T^* is densely defined.*

Proof

The operator T is closable if and only if $\overline{\mathsf{G}(T)}$ does not contain non-zero vertical vectors, which is equivalent to $\begin{bmatrix} 0 & \mathbb{1} \\ -\mathbb{1} & 0 \end{bmatrix} \overline{\mathsf{G}(T)}$ not containing non-zero horizontal vectors. However,

$$\begin{bmatrix} 0 & \mathbb{1} \\ -\mathbb{1} & 0 \end{bmatrix} \overline{\mathsf{G}(T)} = \mathsf{G}(T^*)^{\perp},$$

so $\begin{bmatrix} 0 & \mathbb{1} \\ -\mathbb{1} & 0 \end{bmatrix} \overline{\mathsf{G}(T)}$ does not contain non-zero horizontal vectors if and only if $\mathsf{G}(T^*)^{\perp}$ does not contain non-zero horizontal vectors which, by Proposition 8.3, is equivalent to the fact that $\mathsf{D}(T^*)$ is dense.

□

In particular, if T is closable then T^* is densely defined, so there exists the operator $(T^*)^*$ which we usually denote by T^{**}. We have

$$\mathsf{G}(T^{**}) = \begin{bmatrix} 0 & \mathbb{1} \\ -\mathbb{1} & 0 \end{bmatrix} (\mathsf{G}(T^*)^{\perp}) = \begin{bmatrix} 0 & \mathbb{1} \\ -\mathbb{1} & 0 \end{bmatrix} \left(\begin{bmatrix} 0 & \mathbb{1} \\ -\mathbb{1} & 0 \end{bmatrix} (\mathsf{G}(T)^{\perp})^{\perp} \right) = \overline{\mathsf{G}(T)}.$$

Therefore $T^{**} = \overline{T}$.

Formula (8.5) and elementary properties of the orthogonal complement give the following corollary:

Corollary 8.8 *Let S and T be densely defined operators on $\mathcal{H}$ such that $T \subset S$. Then $S^* \subset T^*$.*

It is important to note the crucial difference between the adjoint operator for bounded and unbounded operators. In the former case we can always write

$$\langle \xi \,|\, x\eta \rangle = \left\langle x^*\xi \,\middle|\, \eta \right\rangle,$$

while in the latter, the formula

$$\langle \phi | T\psi \rangle = \left\langle T^*\phi \,\middle|\, \psi \right\rangle$$

can be used only after making sure that $\phi \in \mathsf{D}(T^*)$.

Remark 8.9 Let us bring attention to the following fact: a self-adjoint operator does not have proper symmetric extensions. Indeed: if $T = T^*$, $T \subset S$ and $S \subset S^*$ then $S \subset S^* \subset T^* = T \subset S$, so that $S = T$.

8.3 Algebraic Operations

Algebraic operations on unbounded operators can become rather technically involved. Let T and S be operators on $\mathcal{H}$. Then the sum $S + T$ is defined on the domain

$$\mathsf{D}(S+T) = \mathsf{D}(S) \cap \mathsf{D}(T)$$

and

$$(S+T)\psi = S\psi + T\psi, \qquad \psi \in \mathsf{D}(S+T).$$

Next we define the product (composition) ST setting

$$\mathsf{D}(ST) = \{\psi \in \mathsf{D}(T) \,|\, T\psi \in \mathsf{D}(S)\}$$

and

$$(ST)\psi = S(T\psi), \qquad \psi \in \mathsf{D}(ST).$$

It turns out that the sum or the product of densely defined operators might not be densely defined (see [Kat, Chapter 6 §1.4] or [ReSi$_1$, Section VIII.1]). Also the sum or product of closed operators might fail to be closed (or even closable). Nevertheless, if $x \in \mathsf{B}(\mathcal{H})$ and T is densely defined that, of course, $T + x$ is densely defined. Similarly, when $u \in \mathsf{B}(\mathcal{H})$ then uT is densely defined and so is Tu if u is invertible.

Proposition 8.10 *Let T be a closed operator on $\mathcal{H}$ and let $x \in \mathsf{B}(\mathcal{H})$. Then the operator $T + x$ is closed.*

Proof
Let $(\psi_n)_{n\in\mathbb{N}}$ be a sequence of elements of $\mathsf{D}(T + x) = \mathsf{D}(T)$ such that

$$\psi_n \xrightarrow[n\to\infty]{} \psi \quad \text{and} \quad T\psi_n + x\psi_n \xrightarrow[n\to\infty]{} \phi$$

for some $\psi, \phi \in \mathcal{H}$. Clearly we then have $T\psi_n \xrightarrow[n\to\infty]{} \phi - x\psi$. As T is closed, we have $\psi \in \mathsf{D}(T)$ and $T\psi = \phi - x\psi$. But this means that $\psi \in \mathsf{D}(T + x)$ and $(T + x)\psi = \phi$. In particular $T + x$ is closed. □

Proposition 8.11 *Let T be a closed operator on $\mathcal{H}$ and let $u \in \mathsf{B}(\mathcal{H})$. Then*
(1) *Tu is closed,*
(2) *if u is invertible then uT is closed.*

Proof

Ad (1). Take a sequence $(\psi_n)_{n\in\mathbb{N}}$ of elements of $\mathsf{D}(Tu)$ and assume that

$$\psi_n \xrightarrow[n\to\infty]{} \psi \quad \text{and} \quad Tu\psi_n \xrightarrow[n\to\infty]{} \phi.$$

Let $\psi_n' = u\psi_n$. Then for each n we have $\psi_n' \in \mathsf{D}(T)$, $\psi_n' \xrightarrow[n\to\infty]{} u\psi$ and

$$T\psi_n' \xrightarrow[n\to\infty]{} \phi.$$

As T is closed, it follows that $u\psi \in \mathsf{D}(T)$ and $T(u\psi) = \phi$. In other words $\psi \in \mathsf{D}(Tu)$ and $(Tu)\psi = \phi$ which shows that Tu is closed.

Ad (2). If $(\psi_n)_{n\in\mathbb{N}}$ is a sequence of elements of $\mathsf{D}(uT) = \mathsf{D}(T)$ such that

$$\psi_n \xrightarrow[n\to\infty]{} \psi \quad \text{and} \quad uT\psi_n \xrightarrow[n\to\infty]{} \phi,$$

then the sequence $(T\psi_n)_{n\in\mathbb{N}}$ satisfies

$$T\psi_n = u^{-1}uT\psi_n \xrightarrow[n\to\infty]{} u^{-1}\phi.$$

Thus, by closedness of T, we have $\psi \in \mathsf{D}(T)$ and $T\psi = u^{-1}\phi$. In other words $\psi \in \mathsf{D}(uT)$ and $uT\psi = \phi$, which means that uT is closed. □

Corollary 8.12 *Let T be a closed operator on $\mathcal{H}$ and let $x \in \mathsf{B}(\mathcal{H})$ be such that $x\mathcal{H} \subset \mathsf{D}(T)$. Then Tx is bounded.*

Proof

By Proposition 8.11(1) the operator Tx is closed. Moreover $\mathsf{D}(Tx) = \mathcal{H}$, so by the closed graph theorem Tx is bounded. □

Proposition 8.13 *Let S and T be densely defined operators on $\mathcal{H}$ such that ST is densely defined.*[1] *Then $T^*S^* \subset (ST)^*$.*

Proof

Take $\psi \in \mathsf{D}(T^*S^*)$. Then for any $\xi \in \mathsf{D}(ST)$ we have

$$\langle \psi | ST\xi \rangle = \langle S^*\psi | T\xi \rangle = \langle T^*S^*\psi | \xi \rangle,$$

since $\psi \in \mathsf{D}(S^*)$ and $S^*\psi \in \mathsf{D}(T^*)$. The right hand side of the above equality is continuous with respect to ξ, and so $\psi \in \mathsf{D}\big((ST)^*\big)$ and $(ST)^*\psi = T^*S^*\psi$. □

[1] Note that it does not follow from density of $\mathsf{D}(ST)$ that the domain of S is dense. Consider e.g. $T = 0$.

Proposition 8.14 *Let T be a densely defined operator on $\mathcal{H}$ and let $x \in \mathsf{B}(\mathcal{H})$. Then $(xT)^* = T^*x^*$.*

Proof
We already know that $T^*x^* \subset (xT)^*$. Take $\eta \in \mathsf{D}\big((xT)^*\big)$. Then for any $\xi \in \mathsf{D}(T) = \mathsf{D}(xT)$ we have

$$\big\langle (xT)^*\eta \big| \xi \big\rangle = \langle \eta | (xT)\xi \rangle = \langle \eta | x(T\xi) \rangle = \big\langle x^*\eta \big| T\xi \big\rangle,$$

i.e. $x^*\eta \in \mathsf{D}(T^*)$ and $T^*(x^*\eta) = (xT)^*\eta$. In other words $\eta \in \mathsf{D}(T^*x^*)$ and $(T^*x^*)\eta = (xT)^*\eta$, which means that $(xT)^* \subset T^*x^*$. □

We also have additive analogs of Propositions 8.13 and 8.14:

Proposition 8.15
(1) *Let T and S be operators such that $T + S$ is densely defined. Then T and S are densely defined and $T^* + S^* \subset (T + S)^*$.*
(2) *Let T be densely defined and let $x \in \mathsf{B}(\mathcal{H})$. Then $(T + x)^* = T^* + x^*$.*

Proof
Ad (1). Clearly if $\mathsf{D}(T + S) = \mathsf{D}(T) \cap \mathsf{D}(S)$ is dense in $\mathcal{H}$ then so are $\mathsf{D}(T)$ and $\mathsf{D}(S)$. Let $\phi \in \mathsf{D}(T^* + S^*) = \mathsf{D}(T^*) \cap \mathsf{D}(S^*)$. Then for any $\psi \in \mathsf{D}(T + S)$ we have

$$\begin{aligned}\langle \phi | (T + S)\psi \rangle = \langle \phi | T\psi \rangle + \langle \phi | S\psi \rangle &= \big\langle T^*\phi \big| \psi \big\rangle + \big\langle S^*\phi \big| \psi \big\rangle \\ &= \big\langle T^*\phi + S^*\phi \big| \psi \big\rangle = \big\langle (T^* + S^*)\phi \big| \psi \big\rangle,\end{aligned}$$

so $\phi \in \mathsf{D}\big((T + S)^*\big)$ and $(T + S)^*\phi = T^*\phi + S^*\phi$.

Ad (2). We already know that $(T + x)^* \supset T^* + x^*$. Take $\eta \in \mathsf{D}\big((T + x)^*\big)$. Then the functional

$$\mathsf{D}(T + x) = \mathsf{D}(T) \ni \psi \longmapsto \langle \eta | (T + x)\psi \rangle \in \mathbb{C}$$

is continuous, and consequently so is the functional

$$\mathsf{D}(T) \ni \psi \longmapsto \langle \eta | T\psi \rangle = \langle \eta | (T + x)\psi \rangle - \langle \eta | x\psi \rangle \in \mathbb{C}$$

In other words $\eta \in \mathsf{D}(T^*)$. Using this fact we can compute

$$\begin{aligned}\langle \eta | (T + x)\psi \rangle = \langle \eta | T\psi \rangle + \langle \eta | x\psi \rangle &= \big\langle T^*\eta \big| \psi \big\rangle + \big\langle x^*\eta \big| \psi \big\rangle \\ &= \big\langle T^*\eta + x^*\eta \big| \psi \big\rangle = \big\langle (T^* + x^*)\eta \big| \psi \big\rangle,\end{aligned}$$

which shows that $(T + x)^* \subset T^* + x^*$. □

8.4 Spectrum

Let T be a closed densely defined operator on $\mathcal{H}$. We say that T is *invertible* if T is a bijection of $\mathsf{D}(T)$ onto $\mathcal{H}$. By the closed graph theorem the inverse bijection $T^{-1} : \mathcal{H} \to \mathsf{D}(T)$ is then bounded. The *spectrum* of the operator T is defined in the same way as for bounded operators:

$$\sigma(T) = \{\lambda \in \mathbb{C} \,|\, \lambda \mathbb{1} - T \text{ is not invertible}\}.$$

Just as in the case of bounded operators, we put $\rho(T) = \mathbb{C} \setminus \sigma(T)$ and call this set the *resolvent set* of T.

Proposition 8.16 *Let T be a closed densely defined operator on $\mathcal{H}$. Then $\sigma(T)$ is a closed subset of $\mathbb{C}$.*

Proof
Take $\lambda_0 \in \rho(T)$. Then for any $\lambda \in \mathbb{C}$ such that

$$|\lambda - \lambda_0| < \left\|(\lambda_0 \mathbb{1} - T)^{-1}\right\|^{-1}$$

the series

$$\sum_{n=0}^{\infty} (\lambda_0 - \lambda)^n (\lambda_0 \mathbb{1} - T)^{-n-1}$$

converges in $\mathsf{B}(\mathcal{H})$ to a sum r. We will now check that $r = (\lambda \mathbb{1} - T)^{-1}$, i.e.
(1) for $\psi \in \mathsf{D}(\lambda \mathbb{1} - T) = \mathsf{D}(T)$ we have $r(\lambda \mathbb{1} - T)\psi = \psi$,
(2) for $\xi \in \mathcal{H}$ we have $r\xi \in \mathsf{D}(T)$ and $(\lambda \mathbb{1} - T)r\xi = \xi$.

Ad (1). Let $\psi \in \mathsf{D}(T)$. Then

$$(\lambda \mathbb{1} - T)\psi = (\lambda - \lambda_0)\psi + (\lambda_0 \mathbb{1} - T)\psi$$

and consequently

$$r(\lambda \mathbb{1} - T)\psi = -\sum_{n=0}^{\infty} (\lambda - \lambda_0)^{n+1} (\lambda_0 \mathbb{1} - T)^{-n-1}\psi + \sum_{n=0}^{\infty} (\lambda - \lambda_0)^n (\lambda_0 \mathbb{1} - T)^{-n}\psi = \psi.$$

Ad (2). For any $n \in \mathbb{Z}_+$ we have $(\lambda_0 \mathbb{1} - T)^{-n-1}\xi \in \mathsf{D}(T)$, so putting

$$\xi_N = \sum_{n=0}^{N} (\lambda_0 - \lambda)^n (\lambda_0 \mathbb{1} - T)^{-n-1}\xi,$$

we obtain a sequence $(\xi_N)_{N\in\mathbb{N}}$ of elements of $\mathsf{D}(\lambda\mathbb{1}-T)$ converging to $r\xi$. Moreover

$$\begin{aligned}
(\lambda\mathbb{1}-T)\xi_N &= \big((\lambda-\lambda_0)\mathbb{1}+(\lambda_0\mathbb{1}-T)\big)\xi_N\\
&= (\lambda-\lambda_0)\xi_N+\sum_{n=0}^{N}(\lambda_0-\lambda)^n(\lambda_0\mathbb{1}-T)^{-n}\xi\\
&= (\lambda-\lambda_0)\xi_N+(\lambda_0-\lambda)\sum_{n=0}^{N}(\lambda_0-\lambda)^{n-1}(\lambda_0\mathbb{1}-T)^{-n}\xi\\
&= (\lambda-\lambda_0)\xi_N\\
&\qquad+(\lambda_0-\lambda)\Big(\tfrac{1}{\lambda_0-\lambda}\xi+\sum_{k=0}^{N-1}(\lambda_0-\lambda)^k(\lambda_0\mathbb{1}-T)^{-k-1}\xi\Big)\\
&= (\lambda-\lambda_0)\xi_N+(\lambda_0-\lambda)\Big(\tfrac{1}{\lambda_0-\lambda}\xi+\xi_{N-1}\Big)\\
&= \xi+(\lambda-\lambda_0)(\xi_N-\xi_{N-1})\xrightarrow[N\to\infty]{}\xi.
\end{aligned}$$

Thus, by closedness of $(\lambda\mathbb{1}-T)$, we have $r\xi\in\mathsf{D}(\lambda\mathbb{1}-T)$ and $(\lambda\mathbb{1}-T)r\xi=\xi$.

This way we showed that $(\lambda\mathbb{1}-T)$ is invertible, so $\lambda\in\rho(T)$. It follows that $\rho(T)$ is an open subset of $\mathbb{C}$, an therefore $\sigma(T)=\mathbb{C}\setminus\rho(T)$ is closed. □

It is worth mentioning that one can construct examples of closed densely defined operators T with $\sigma(T)=\emptyset$, as well as operators whose spectrum is all of $\mathbb{C}$ (see e.g. [ReSi$_1$, Example 5, p. 254]). It is also not hard to see that if u is unitary, then $\sigma(uTu^*)=\sigma(T)$. Finally let us add that some authors prefer a slightly different definition of the spectrum. In this other approach the spectrum of T is considered as a subset of the Riemann sphere $\overline{\mathbb{C}}$ and by definition contains the point ∞ whenever T is not bounded. This version of the spectrum is sometimes called the *extended spectrum*. One of the benefits of considering the extended spectrum is that it is always a non-empty and compact subset of $\overline{\mathbb{C}}$, regardless whether the considered operator is bounded or not.

Notes

Fundamentals of the theory of unbounded operators on a Hilbert space presented above are developed further in monographs such as [AkGl, Kat, Mau, ReSi$_1$, ReSi$_2$]. Moreover, many textbooks of general functional analysis have separate sections dealing with unbounded operators (e.g. [Ped, Chapter 5], [Rud$_2$, Chapter 13]). Our intention in this chapter was to introduce the reader to the topic reserving a more in-depth analysis of several aspects of the theory for the following chapters. Many examples and exercises can be found in textbooks and monographs mentioned above.

z-Transform

P. Sołtan, *A Primer on Hilbert Space Operators*, Compact Textbooks in Mathematics,
https://doi.org/10.1007/978-3-319-92061-0_9

This chapter will be devoted to developing an extremely useful tool for dealing with unbounded operators, namely the so called z-transform. It was introduced in a context much wider than the theory of operators on Hilbert spaces by S.L. Woronowicz (see [Lan, WoNa]). As we already mentioned a couple of times, the z-transform is a way to encode full information about a given closed densely defined operator T in a bounded operator z_T. The procedure of passing from T to z_T requires several preliminary results which will be presented in ▶ Sect. 9.1. As an illustration of the use of the z-transform we will provide a simple proof of existence of polar decomposition of closed operators given in ▶ Sect. 9.3.

9.1 The Operator T^*T

Let T be a closed operator on $\mathcal{H}$. A subspace $\mathscr{D} \subset \mathsf{D}(T)$ is called a *core* for T if T is equal to the closure of the restriction $T\big|_{\mathscr{D}}$ of T to $\mathscr{D}$.

From now on assume that T is closed and densely defined. Recall that

$$\mathsf{G}(T)^{\perp} = \begin{bmatrix} 0 & \mathbb{1} \\ -\mathbb{1} & 0 \end{bmatrix} \mathsf{G}(T^*) = \left\{ \begin{bmatrix} T^*\phi \\ -\phi \end{bmatrix} \,\middle|\, \phi \in \mathsf{D}(T^*) \right\}.$$

It follows that for any $\left[\begin{smallmatrix} \xi \\ \eta \end{smallmatrix}\right] \in \mathcal{H} \oplus \mathcal{H}$ there exists $\psi \in \mathsf{D}(T)$ and $\phi \in \mathsf{D}(T^*)$ such that

$$\begin{bmatrix} \xi \\ \eta \end{bmatrix} = \begin{bmatrix} \psi \\ T\psi \end{bmatrix} + \begin{bmatrix} T^*\phi \\ -\phi \end{bmatrix}. \tag{9.1}$$

Proposition 9.1 *For any $\xi \in \mathcal{H}$ there exists a unique $\psi \in \mathsf{D}(T^*T)$ such that $\xi = (\mathbb{1} + T^*T)\psi$. Moreover $\|\psi\| \leq \|\xi\|$.*

Proof

Put $\eta = 0$ in (9.1). Then there exist $\psi \in \mathsf{D}(T)$ and $\phi \in \mathsf{D}(T^*)$ such that

$$\begin{bmatrix} \xi \\ 0 \end{bmatrix} = \begin{bmatrix} \psi \\ T\psi \end{bmatrix} + \begin{bmatrix} T^*\phi \\ -\phi \end{bmatrix}$$

which means precisely that $\xi = \psi + T^*\phi$ and $\phi = T\psi$. In other words $\psi \in \mathsf{D}(T^*T)$ and $\psi + T^*T\psi = \xi$.

We have

$$\begin{aligned} \|\xi\|^2 &= \langle \psi + T^*T\psi \,|\, \psi + T^*T\psi \rangle \\ &= \langle \psi \,|\, \psi \rangle + \langle T^*T\psi \,|\, \psi \rangle + \langle \psi \,|\, T^*T\psi \rangle + \langle T^*T\psi \,|\, T^*T\psi \rangle \\ &= \|\psi\|^2 + 2\|T\psi\|^2 + \|T^*T\psi\|^2 \geq \|\psi\|^2. \end{aligned}$$

The above estimate shows also that the vector ψ is uniquely determined by ξ. Indeed: if $\xi = \psi + T^*T\psi = \psi' + T^*T\psi'$ then

$$0 = (\psi - \psi') + T^*T(\psi - \psi')$$

and we obtain $\|\psi - \psi'\| \leq 0$, so that $\psi = \psi'$. □

It follows from Proposition 9.1 that

- $\mathsf{D}(T^*T) \neq \{0\}$,
- the map

$$\mathsf{D}(T^*T) \ni \psi \longmapsto \psi + T^*T\psi \in \mathcal{H} \tag{9.2}$$

is a bijection which does not decrease the norm.

In particular, the map $(\mathbb{1} + T^*T)^{-1}$, i.e. the inverse of (9.2) is continuous and $\left\|(\mathbb{1} + T^*T)^{-1}\right\| \leq 1$.

Theorem 9.2

Let T be a closed densely defined operator. Then

(1) *the operator T^*T is closed,*

(2) $\mathsf{D}(T^*T)$ *is a core for T.*

Proof
We begin with Statement (2). Suppose $\mathsf{G}(T|_{\mathsf{D}(T^*T)})$ is not dense in $\mathsf{G}(T)$. This means that there exists a non-zero $\phi \in \mathsf{D}(T)$ such that the vector $\left[\begin{smallmatrix}\phi \\ T\phi\end{smallmatrix}\right]$ is orthogonal to $\mathsf{G}(T|_{\mathsf{D}(T^*T)})$, i.e.

$$\begin{bmatrix}\phi \\ T\phi\end{bmatrix} \perp \begin{bmatrix}\psi \\ T\psi\end{bmatrix}, \qquad \psi \in \mathsf{D}(T^*T)$$

or, in other words,

$$\langle\phi|\psi\rangle + \langle T\phi|T\psi\rangle = 0, \qquad \psi \in \mathsf{D}(T^*T).$$

This, however, means that

$$\langle\phi|\psi + T^*T\psi\rangle = 0, \qquad \psi \in \mathsf{D}(T^*T),$$

so that $\phi \perp \mathcal{H}$. This contradiction shows that $\mathsf{D}(T^*T)$ is a core for T.

Statement (1) is proved as follows: the operator $(\mathbb{1} + T^*T)^{-1}$ is bounded and consequently its graph is closed. Therefore the graph of $\mathbb{1} + T^*T$ is closed, i.e. $\mathbb{1} + T^*T$ is closed. Finally $T^*T = (\mathbb{1} + T^*T) + (-\mathbb{1})$ is closed by Proposition 8.10. □

Corollary 9.3 *Let T be a closed densely defined operator on $\mathcal{H}$. Then $\mathsf{D}(T^*T)$ is dense in $\mathcal{H}$.*

Proof
A core of an operator is dense in its domain and T is assumed to be densely defined. □

9.2 z-Transform of a Closed Operator

Throughout this section T will be a closed densely defined operator on $\mathcal{H}$.

Lemma 9.4 *The operator $(\mathbb{1} + T^*T)^{-1}$ is positive.*

Proof
Take $\xi \in \mathcal{H}$ and let $\psi \in \mathsf{D}(T^*T)$ be such that $\xi = \psi + T^*T\psi$. Then

$$\begin{aligned}\left\langle\xi\middle|(\mathbb{1} + T^*T)^{-1}\xi\right\rangle &= \left\langle\psi + T^*T\psi\middle|(\mathbb{1} + T^*T)^{-1}(\mathbb{1} + T^*T)\psi\right\rangle \\ &= \langle\psi + T^*T\psi|\psi\rangle = \langle\psi|\psi\rangle + \langle T^*T\psi|\psi\rangle \\ &= \|\psi\|^2 + \|T\psi\|^2 \geq 0.\end{aligned}$$

□

Lemma 9.4 allows us to consider the operator $(\mathbb{1}+T^*T)^{-\frac{1}{2}}$, i.e. the positive square root of the positive operator $(\mathbb{1}+T^*T)^{-1}$. Note that its range is dense in $\mathcal{H}$, as

$$\begin{aligned}(\mathbb{1}+T^*T)^{-\frac{1}{2}}\mathcal{H} &\supset (\mathbb{1}+T^*T)^{-\frac{1}{2}}(\mathbb{1}+T^*T)^{-\frac{1}{2}}\mathcal{H}\\ &= (\mathbb{1}+T^*T)^{-1}\mathcal{H} = \mathsf{D}(T^*T)\end{aligned}$$

and the operator T^*T is densely defined.

Theorem 9.5

Let T be a closed densely defined operator on $\mathcal{H}$. Then
(1) $(\mathbb{1}+T^*T)^{-\frac{1}{2}}\mathcal{H} = \mathsf{D}(T)$,
(2) $T(\mathbb{1}+T^*T)^{-\frac{1}{2}} \in \mathsf{B}(\mathcal{H})$ *and* $\left\|T(\mathbb{1}+T^*T)^{-\frac{1}{2}}\right\| \le 1$.

Proof

Take $\eta \in \mathcal{H}$. Then $(\mathbb{1}+T^*T)^{-1}\eta \in \mathsf{D}(T^*T) \subset \mathsf{D}(T)$ and

$$\begin{aligned}\left\|T(\mathbb{1}+T^*T)^{-1}\eta\right\|^2 &= \left\langle T(\mathbb{1}+T^*T)^{-1}\eta \middle| T(\mathbb{1}+T^*T)^{-1}\eta\right\rangle\\ &= \left\langle (\mathbb{1}+T^*T)^{-1}\eta \middle| T^*T(\mathbb{1}+T^*T)^{-1}\eta\right\rangle\\ &\le \left\langle (\mathbb{1}+T^*T)^{-1}\eta \middle| (\mathbb{1}+T^*T)(\mathbb{1}+T^*T)^{-1}\eta\right\rangle\\ &= \left\langle (\mathbb{1}+T^*T)^{-1}\eta \middle| \eta\right\rangle\\ &= \left\langle (\mathbb{1}+T^*T)^{-\frac{1}{2}}\eta \middle| (\mathbb{1}+T^*T)^{-\frac{1}{2}}\eta\right\rangle = \left\|(\mathbb{1}+T^*T)^{-\frac{1}{2}}\eta\right\|^2.\end{aligned}$$

Therefore, for vectors ξ of the form $(\mathbb{1}+T^*T)^{-\frac{1}{2}}\eta$ we have

$$\left\|T(\mathbb{1}+T^*T)^{-\frac{1}{2}}\xi\right\| \le \|\xi\|. \tag{9.3}$$

Now let $\zeta \in \mathcal{H}$. Then there exists a sequence $(\xi_n)_{n\in\mathbb{N}}$ of elements of the subspace $(\mathbb{1}+T^*T)^{-\frac{1}{2}}\mathcal{H}$ converging to ζ. It follows that

$$(\mathbb{1}+T^*T)^{-\frac{1}{2}}\xi_n \xrightarrow[n\to\infty]{} (\mathbb{1}+T^*T)^{-\frac{1}{2}}\zeta.$$

Moreover, thanks to the estimate (9.3), we also have

$$\begin{aligned}&\left\|T(\mathbb{1}+T^*T)^{-\frac{1}{2}}\xi_n - T(\mathbb{1}+T^*T)^{-\frac{1}{2}}\xi_m\right\|\\ &\qquad= \left\|T(\mathbb{1}+T^*T)^{-\frac{1}{2}}(\xi_n-\xi_m)\right\| \le \|\xi_n-\xi_m\|,\end{aligned}$$

which implies that the sequence $\left(T(\mathbb{1} + T^*T)^{-\frac{1}{2}}\xi_n\right)_{n\in\mathbb{N}}$ converges. This way, form closedness of T we infer that $(\mathbb{1} + T^*T)^{-\frac{1}{2}}\zeta \in \mathsf{D}(T)$ and consequently

$$(\mathbb{1} + T^*T)^{-\frac{1}{2}}\mathcal{H} \subset \mathsf{D}(T).$$

By Corollary 8.12, this shows that $T(\mathbb{1} + T^*T)^{-\frac{1}{2}} \in \mathsf{B}(\mathcal{H})$ and (9.3) gives $\left\|T(\mathbb{1} + T^*T)^{-\frac{1}{2}}\right\| \leq 1$. Statement (2) is proved.

To finish the proof of (1) we have to show that $\mathsf{D}(T) \subset (\mathbb{1} + T^*T)^{-\frac{1}{2}}\mathcal{H}$. Let us take any $\xi \in \mathsf{D}(T)$. Since $\mathsf{D}(T^*T)$ is a core for T, there exists a sequence $(\psi_n)_{n\in\mathbb{N}}$ of elements of $\mathsf{D}(T^*T)$ such that

$$\psi_n \xrightarrow[n\to\infty]{} \xi \quad \text{and} \quad T\psi_n \xrightarrow[n\to\infty]{} T\xi.$$

Put $\phi_n = (\mathbb{1} + T^*T)\psi_n$. Then the sequence $\left((\mathbb{1} + T^*T)^{-\frac{1}{2}}\phi_n\right)_{n\in\mathbb{N}}$ converges, as

$$\begin{aligned}
&\left\|(\mathbb{1} + T^*T)^{-\frac{1}{2}}(\phi_n - \phi_m)\right\|^2 \\
&\qquad = \left\langle(\mathbb{1} + T^*T)^{-\frac{1}{2}}(\phi_n - \phi_m)\,\middle|\,(\mathbb{1} + T^*T)^{-\frac{1}{2}}(\phi_n - \phi_m)\right\rangle \\
&\qquad = \left\langle\phi_n - \phi_m\,\middle|\,(\mathbb{1} + T^*T)^{-1}(\phi_n - \phi_m)\right\rangle \\
&\qquad = \left\langle(\mathbb{1} + T^*T)(\psi_n - \psi_m)\,\middle|\,\psi_n - \psi_m\right\rangle \\
&\qquad = \|\psi_n - \psi_m\|^2 + \|T\psi_n - T\psi_m\|^2 \xrightarrow[n,m\to\infty]{} 0.
\end{aligned}$$

It follows that

$$\begin{aligned}
\xi = \lim_{n\to\infty}\psi_n &= \lim_{n\to\infty}(\mathbb{1} + T^*T)^{-\frac{1}{2}}(\mathbb{1} + T^*T)^{-\frac{1}{2}}\phi_n \\
&= (\mathbb{1} + T^*T)^{-\frac{1}{2}}\lim_{n\to\infty}(\mathbb{1} + T^*T)^{-\frac{1}{2}}\phi_n \in (\mathbb{1} + T^*T)^{-\frac{1}{2}}\mathcal{H}.
\end{aligned}$$

□

The bounded operator $T(\mathbb{1} + T^*T)^{-\frac{1}{2}}$ is called the *z-transform* of T and is denoted by the symbol z_T. Let us note once more that for any closed densely defined T we have $\|z_T\| \leq 1$. In particular $z_T{}^*z_T \leq \mathbb{1}$, so that the operator $\mathbb{1} - z_T{}^*z_T$ is positive.

The next theorem makes precise the claim that all information about T is contained in z_T.

Theorem 9.6
Let T be a closed densely defined operator on $\mathcal{H}$. Then

$$\mathsf{G}(T) = \left\{ \begin{bmatrix} (\mathbb{1} - z_T{}^* z_T)^{\frac{1}{2}}\xi \\ z_T \xi \end{bmatrix} \middle| \, \xi \in \mathcal{H} \right\}.$$

Proof
We have

$$\begin{aligned} \mathsf{G}(T) &= \left\{ \begin{bmatrix} \psi \\ T\psi \end{bmatrix} \middle| \, \psi \in \mathsf{D}(T) \right\} \\ &= \left\{ \begin{bmatrix} (\mathbb{1} - T^*T)^{-\frac{1}{2}}\xi \\ T(\mathbb{1} - T^*T)^{-\frac{1}{2}}\xi \end{bmatrix} \middle| \, \xi \in \mathcal{H} \right\} \\ &= \left\{ \begin{bmatrix} (\mathbb{1} - T^*T)^{-\frac{1}{2}}\xi \\ z_T \xi \end{bmatrix} \middle| \, \xi \in \mathcal{H} \right\}. \end{aligned} \tag{9.4}$$

Now, remembering that for any $\xi \in \mathcal{H}$ the vector $(\mathbb{1} + T^*T)^{-1}\xi$ belongs to $\mathsf{D}(T^*T)$, we compute

$$\begin{aligned} \left\|(\mathbb{1} + T^*T)^{-\frac{1}{2}}\xi\right\|^2 &= \left\langle (\mathbb{1} + T^*T)^{-\frac{1}{2}}\xi \middle| (\mathbb{1} + T^*T)^{-\frac{1}{2}}\xi \right\rangle \\ &= \left\langle \xi \middle| (\mathbb{1} + T^*T)^{-1}\xi \right\rangle = \left\langle (\mathbb{1} + T^*T)^{-1}\xi \middle| \xi \right\rangle \\ &= \left\langle (\mathbb{1} + T^*T)^{-1}\xi \middle| (\mathbb{1} + T^*T)(\mathbb{1} + T^*T)^{-1}\xi \right\rangle \\ &= \left\|(\mathbb{1} + T^*T)^{-1}\xi\right\|^2 + \left\langle (\mathbb{1} + T^*T)^{-1}\xi \middle| T^*T(\mathbb{1} + T^*T)^{-1}\xi \right\rangle \\ &= \left\|(\mathbb{1} + T^*T)^{-1}\xi\right\|^2 + \left\|T(\mathbb{1} + T^*T)^{-1}\xi\right\|^2 \\ &= \left\|(\mathbb{1} + T^*T)^{-\frac{1}{2}}(\mathbb{1} + T^*T)^{-\frac{1}{2}}\xi\right\|^2 \\ &\qquad + \left\|T(\mathbb{1} + T^*T)^{-\frac{1}{2}}(\mathbb{1} + T^*T)^{-\frac{1}{2}}\xi\right\|^2. \end{aligned}$$

Therefore, setting $\psi = (\mathbb{1} + T^*T)^{-\frac{1}{2}}\xi$ we get

$$\|\psi\|^2 = \left\|(\mathbb{1} + T^*T)^{-\frac{1}{2}}\psi\right\|^2 + \|z_T\psi\|^2. \tag{9.5}$$

Such vectors ψ form a dense subset $\mathsf{D}(T)$ of $\mathcal{H}$, an consequently (9.5) holds for all $\psi \in \mathcal{H}$.

It follows from (9.5) and the polarization formula that

$$\langle\phi|\psi\rangle = \left\langle\phi\middle|(\mathbb{1}+T^*T)^{-1}\psi\right\rangle + \langle\phi|z_T{}^*z_T\psi\rangle, \qquad \psi,\phi\in\mathcal{H}.$$

This, in turn, means that

$$\psi = (\mathbb{1}+T^*T)^{-1}\psi + z_T{}^*z_T\psi, \qquad \psi\in\mathcal{H},$$

so that

$$(\mathbb{1}+T^*T)^{-1} = \mathbb{1} - z_T{}^*z_T. \tag{9.6}$$

Therefore $(\mathbb{1}+T^*T)^{-\frac{1}{2}} = (\mathbb{1}-z_T{}^*z_T)^{\frac{1}{2}}$ which, together with (9.4), yields

$$\mathsf{G}(T) = \left\{\begin{bmatrix}(\mathbb{1}-z_T{}^*z_T)^{\frac{1}{2}}\xi\\ z_T\xi\end{bmatrix}\middle|\ \xi\in\mathcal{H}\right\}.$$

□

Corollary 9.7 *Let S and T be closed densely defined operators on $\mathcal{H}$. If $z_S = z_T$ then $S = T$.*

Let us note here that if T is a closed densely defined operator then z_T satisfies

$$\ker(\mathbb{1}-z_T{}^*z_T) = \{0\}.$$

Indeed: $\ker(\mathbb{1}-z_T{}^*z_T) = \big((\mathbb{1}-z_T{}^*z_T)\mathcal{H}\big)^\perp$ and by (9.6) we have

$$(\mathbb{1}-z_T{}^*z_T)\mathcal{H} = (\mathbb{1}+T^*T)^{-1}\mathcal{H} = \mathsf{D}(T^*T),$$

so that $\big((\mathbb{1}-z_T{}^*z_T)\mathcal{H}\big)^\perp = \{0\}$.

Lemma 9.8 *Let $z\in\mathsf{B}(\mathcal{H})$ be such that $\|z\|\le 1$. Then for any $f\in\mathrm{C}([0,1])$ we have*

$$f(z^*z)z^* = z^*f(zz^*) \quad \textit{and} \quad zf(z^*z) = f(zz^*)z.$$

Proof

We have $\sigma(z^*z),\sigma(zz^*)\subset[0,1]$. Let $(f_n)_{n\in\mathbb{N}}$ be a sequence of polynomials converging uniformly to f on $[0,1]$. Then

$$f_n(z^*z)z^* = z^*f_n(zz^*), \qquad n\in\mathbb{N}$$

and

$$f_n(z^*z)z^* \xrightarrow[n\to\infty]{} f(z^*z)z^*, \quad z^*f_n(zz^*) \xrightarrow[n\to\infty]{} z^*f(zz^*).$$

It follows that $f(z^*z)z^* = z^*f(zz^*)$. The second formula is obtained from the first one for $\overline{f}$ by applying the involution. □

Theorem 9.9
An operator $z \in \mathsf{B}(\mathcal{H})$ is a z-transform of a closed densely defined operator T if and only if z satisfies

$$\|z\| \leq 1 \quad \text{and} \quad \ker(\mathbb{1} - z^*z) = \{0\}.$$

Proof
We checked above that the conditions of the theorem are necessary for z to be a z-transform of a closed densely defined operator. Assume now that $z \in \mathsf{B}(\mathcal{H})$ satisfies $\|z\| \leq 1$ and $\ker(\mathbb{1} - z^*z) = \{0\}$. Put

$$\mathsf{G} = \left\{ \begin{bmatrix} (\mathbb{1} - z^*z)^{\frac{1}{2}}\xi \\ z\xi \end{bmatrix} \,\middle|\, \xi \in \mathcal{H} \right\}.$$

We will check that G is a graph of a closed densely defined operator (see Corollary 8.4). To that end let us define

$$U_z = \begin{bmatrix} (\mathbb{1} - z^*z)^{\frac{1}{2}} & -z^* \\ z & (\mathbb{1} - zz^*)^{\frac{1}{2}} \end{bmatrix} \in \mathsf{B}(\mathcal{H} \oplus \mathcal{H}).$$

Then U_z is unitary:

$$\begin{aligned} U_z^*U_z &= \begin{bmatrix} (\mathbb{1} - z^*z)^{\frac{1}{2}} & z^* \\ -z & (\mathbb{1} - zz^*)^{\frac{1}{2}} \end{bmatrix} \begin{bmatrix} (\mathbb{1} - z^*z)^{\frac{1}{2}} & -z^* \\ z & (\mathbb{1} - zz^*)^{\frac{1}{2}} \end{bmatrix} \\ &= \begin{bmatrix} (\mathbb{1} - z^*z) + z^*z & z^*(\mathbb{1} - zz^*)^{\frac{1}{2}} - (\mathbb{1} - z^*z)^{\frac{1}{2}}z^* \\ (\mathbb{1} - zz^*)^{\frac{1}{2}}z - z(\mathbb{1} - z^*z)^{\frac{1}{2}} & zz^* + (\mathbb{1} - zz^*) \end{bmatrix} \\ &= \begin{bmatrix} \mathbb{1} & 0 \\ 0 & \mathbb{1} \end{bmatrix} \end{aligned}$$

by Lemma 9.8 applied to the function $f(t) = \sqrt{1-t}$. Similarly

$$U_z U_z^* = \begin{bmatrix} (\mathbb{1} - z^*z)^{\frac{1}{2}} & -z^* \\ z & (\mathbb{1} - zz^*)^{\frac{1}{2}} \end{bmatrix} \begin{bmatrix} (\mathbb{1} - z^*z)^{\frac{1}{2}} & z^* \\ -z & (\mathbb{1} - zz^*)^{\frac{1}{2}} \end{bmatrix}$$

$$= \begin{bmatrix} (\mathbb{1} - z^*z) + z^*z & (\mathbb{1} - z^*z)^{\frac{1}{2}} z^* - z^*(\mathbb{1} - zz^*)^{\frac{1}{2}} \\ z(\mathbb{1} - z^*z)^{\frac{1}{2}} - (\mathbb{1} - zz^*)^{\frac{1}{2}} z & zz^* + (\mathbb{1} - zz^*) \end{bmatrix}$$

$$= \begin{bmatrix} \mathbb{1} & 0 \\ 0 & \mathbb{1} \end{bmatrix}.$$

Note further that

$$\mathsf{G} = U_z \left\{ \begin{bmatrix} \xi \\ 0 \end{bmatrix} \middle| \, \xi \in \mathcal{H} \right\},$$

so that G is a closed subspace of $\mathcal{H} \oplus \mathcal{H}$ (as the image of one under a unitary map). We check now that G does not contain non-zero vertical vectors: suppose $\left[\begin{smallmatrix} 0 \\ \phi \end{smallmatrix}\right] \in \mathsf{G}$. Then there exists $\xi \in \mathcal{H}$ such that

$$\begin{bmatrix} 0 \\ \phi \end{bmatrix} = \begin{bmatrix} (\mathbb{1} - z^*z)^{\frac{1}{2}} \xi \\ z\xi \end{bmatrix}.$$

This means that $(\mathbb{1} - z^*z)^{\frac{1}{2}} \xi = 0$, but this implies $(\mathbb{1} - z^*z)\xi = 0$, so $\xi \in \ker(\mathbb{1} - z^*z) = \{0\}$.

To see that $\mathsf{G}^{\perp}$ does not contain non-zero horizontal vectors we note that

$$\mathsf{G}^{\perp} = \left(U_z \left\{ \begin{bmatrix} \xi \\ 0 \end{bmatrix} \middle| \, \xi \in \mathcal{H} \right\} \right)^{\perp}$$

$$= U_z \left(\left\{ \begin{bmatrix} \xi \\ 0 \end{bmatrix} \middle| \, \xi \in \mathcal{H} \right\}^{\perp} \right) = U_z \left\{ \begin{bmatrix} 0 \\ \eta \end{bmatrix} \middle| \, \eta \in \mathcal{H} \right\}$$

$$= \left\{ \begin{bmatrix} -z^*\eta \\ (\mathbb{1} - zz^*)^{\frac{1}{2}} \eta \end{bmatrix} \middle| \, \eta \in \mathcal{H} \right\}.$$

Suppose that $\left[\begin{smallmatrix} \phi \\ 0 \end{smallmatrix}\right] \in \mathsf{G}^{\perp}$. Then there exists $\eta \in \mathcal{H}$ such that

$$\begin{bmatrix} \phi \\ 0 \end{bmatrix} = \begin{bmatrix} -z^*\eta \\ (\mathbb{1} - zz^*)^{\frac{1}{2}} \eta \end{bmatrix}.$$

Therefore $(\mathbb{1} - zz^*)^{\frac{1}{2}} \eta = 0$, so also $(\mathbb{1} - zz^*)\eta = 0$. In other words

$$zz^*\eta = \eta.$$

Multiplying this equality from the left by z^* gives $z^*\eta \in \ker(\mathbb{1} - z^*z) = \{0\}$, so $\phi = -z^*\eta = 0$.
□

Repeating the reasoning of the proof of Theorem 9.9 we obtain the following fact: for a closed densely defined operator T on $\mathcal{H}$ we have

$$\mathsf{G}(T)^{\perp} = \left\{ \begin{bmatrix} -z_T{}^*\xi \\ (\mathbb{1} - z_T z_T{}^*)^{\frac{1}{2}}\xi \end{bmatrix} \middle| \xi \in \mathcal{H} \right\}. \tag{9.7}$$

Indeed:

$$\mathsf{G}(T) = \left\{ \begin{bmatrix} \psi \\ T\psi \end{bmatrix} \middle| \psi \in \mathsf{D}(T) \right\} = U_{z_T}\big(\mathcal{H} \oplus \{0\}\big),$$

where

$$U_{z_T} = \begin{bmatrix} (\mathbb{1} - z_T{}^* z_T)^{\frac{1}{2}} & -z_T{}^* \\ z_T & (\mathbb{1} - z_T z_T{}^*)^{\frac{1}{2}} \end{bmatrix},$$

and $\mathcal{H} \oplus \{0\} = \{[\begin{smallmatrix} \xi \\ 0 \end{smallmatrix}] \mid \xi \in \mathcal{H}\}$. Since U_{z_T} is unitary (the argument for that is the calculation from the proof of Theorem 9.9), we obtain

$$\mathsf{G}(T)^{\perp} = \Big(U_{z_T}\big(\mathcal{H} \oplus \{0\}\big)\Big)^{\perp} = U_{z_T}\Big(\big(\mathcal{H} \oplus \{0\}\big)^{\perp}\Big) = U_{z_T}\big(\{0\} \oplus \mathcal{H}\big),$$

where $\{0\} \oplus \mathcal{H}$ denotes $\{[\begin{smallmatrix} 0 \\ \eta \end{smallmatrix}] \,|\, \xi \in \mathcal{H}\}$. This immediately gives (9.7).

Proposition 9.10 *Let T be closed densely defined operator on $\mathcal{H}$. Then $z_{T^*} = z_T{}^*$.*

Proof
Using formula (9.7) we get

$$\begin{aligned} \mathsf{G}(T^*) &= \begin{bmatrix} 0 & \mathbb{1} \\ -\mathbb{1} & 0 \end{bmatrix} \mathsf{G}(T)^{\perp} \\ &= \begin{bmatrix} 0 & \mathbb{1} \\ -\mathbb{1} & 0 \end{bmatrix} \left\{ \begin{bmatrix} -z_T{}^*\xi \\ (\mathbb{1} - z_T z_T{}^*)^{\frac{1}{2}}\xi \end{bmatrix} \middle| \xi \in \mathcal{H} \right\} \\ &= \left\{ \begin{bmatrix} (\mathbb{1} - z_T z_T{}^*)^{\frac{1}{2}}\xi \\ z_T{}^*\xi \end{bmatrix} \middle| \xi \in \mathcal{H} \right\}. \end{aligned}$$

It follows that $z_T{}^*$ is the z-transform of T^*.
□

We end this section with a remark on how the z-transform behaves under unitary equivalence:

Remark 9.11 Let $\mathcal{H}$ and $\mathcal{K}$ be Hilbert spaces and let T be a closed densely defined operator on $\mathcal{H}$. Let $u \in \mathsf{B}(\mathcal{H}, \mathcal{K})$ be a unitary operator and let $S = u^*Tu$, i.e.

$$\mathsf{D}(S) = \{\psi \in \mathcal{K} \,|\, u\psi \in \mathsf{D}(T)\} = \{u^*\eta \,|\, \eta \in \mathsf{D}(T)\}$$

and $S(u^*\eta) = u^*T\eta$ for all $\eta \in \mathsf{D}(T)$. Then S is a closed and densely defined operator on $\mathcal{K}$ and $z_S = u^*z_T u$. Indeed:

$$\begin{aligned}
\mathsf{G}(S) &= \left\{ \begin{bmatrix} \psi \\ S\psi \end{bmatrix} \middle| \, \psi \in \mathsf{D}(S) \right\} \\
&= \left\{ \begin{bmatrix} u^*\eta \\ u^*T\eta \end{bmatrix} \middle| \, \eta \in \mathsf{D}(T) \right\} \\
&= \left\{ \begin{bmatrix} u^*(\mathbb{1} - z_T{}^* z_T)^{\frac{1}{2}}\xi \\ u^* z_T \xi \end{bmatrix} \middle| \, \xi \in \mathcal{H} \right\} \\
&= \left\{ \begin{bmatrix} u^*(\mathbb{1} - z_T{}^* z_T)^{\frac{1}{2}}u\phi \\ u^* z_T u\phi \end{bmatrix} \middle| \, \phi \in \mathcal{K} \right\} \\
&= \left\{ \begin{bmatrix} (\mathbb{1} - z^* z)^{\frac{1}{2}}\phi \\ z\phi \end{bmatrix} \middle| \, \phi \in \mathcal{K} \right\}
\end{aligned}$$

with $z = u^*z_T u$. In view of Corollary 9.7, this shows that $z_S = u^*z_T u$.

9.3 Polar Decomposition

An operator T on $\mathcal{H}$ is called *positive* if

$$\langle \xi \,|\, T\xi \rangle \geq 0, \qquad \xi \in \mathsf{D}(T).$$

Unlike in the case of bounded operators, positivity of an unbounded operator does not imply its self-adjointness. However, for a closed densely defined operator T the operator T^*T is always positive and self-adjoint.[1]

[1] It is easy to see that T^*T is positive. To see that T^*T is self-adjoint let $a = (\mathbb{1} + T^*T)^{-1}$. Then a is self-adjoint, so $\left[\begin{smallmatrix} 0 & \mathbb{1} \\ -\mathbb{1} & 0 \end{smallmatrix}\right]\mathsf{G}(a)^{\perp} = \mathsf{G}(a)$. Applying the unitary operator $\left[\begin{smallmatrix} 0 & \mathbb{1} \\ \mathbb{1} & 0 \end{smallmatrix}\right]$ to both sides of this equality we obtain

$$\begin{bmatrix} 0 & -\mathbb{1} \\ \mathbb{1} & 0 \end{bmatrix} \mathsf{G}(\mathbb{1} + T^*T)^{\perp} = \mathsf{G}(\mathbb{1} + T^*T),$$

which shows that $S = \mathbb{1} + T^*T$ is self-adjoint. Thus, by Proposition 8.15(2) we get $(T^*T)^* = (S - \mathbb{1})^* = S - \mathbb{1} = T^*T$.

Lemma 9.12 *An operator T is positive if and only if*

$$(\mathbb{1} - z_T{}^* z_T)^{\frac{1}{2}} z_T \geq 0.$$

In particular T is positive and self-adjoint if and only if $z_T \geq 0$.

Proof

A vector ψ belongs to $\mathsf{D}(T)$ if and only if ψ is of the form $(\mathbb{1} - z_T{}^* z_T)^{\frac{1}{2}} \xi$ for some $\xi \in \mathcal{H}$. Therefore

$$\begin{aligned} \langle \psi | T \psi \rangle &= \left\langle (\mathbb{1} - z_T{}^* z_T)^{\frac{1}{2}} \xi \,\middle|\, T(\mathbb{1} - z_T{}^* z_T)^{\frac{1}{2}} \xi \right\rangle \\ &= \left\langle (\mathbb{1} - z_T{}^* z_T)^{\frac{1}{2}} \xi \,\middle|\, z_T \xi \right\rangle = \left\langle \xi \,\middle|\, (\mathbb{1} - z_T{}^* z_T)^{\frac{1}{2}} z_T \xi \right\rangle \end{aligned}$$

which proves the first part of the lemma. Furthermore, $T = T^*$ if and only if $z_T = z_T{}^*$, so T is positive and self-adjoint if and only if $z_T = z_T{}^*$ and

$$\begin{aligned} &\left\langle (\mathbb{1} - z_T{}^* z_T)^{\frac{1}{4}} \eta \,\middle|\, z_T (\mathbb{1} - z_T{}^* z_T)^{\frac{1}{4}} \eta \right\rangle \\ &\qquad = \left\langle \eta \,\middle|\, (\mathbb{1} - z_T{}^* z_T)^{\frac{1}{2}} z_T \eta \right\rangle \geq 0, \qquad \eta \in \mathcal{H}. \end{aligned} \tag{9.8}$$

It is easy to see that the range of $(\mathbb{1} - z_T{}^* z_T)^{\frac{1}{4}}$ is dense in $\mathcal{H}$, and hence the condition (9.8) is equivalent to

$$\langle \varphi | z_T \varphi \rangle \geq 0, \qquad \varphi \in \mathcal{H},$$

i.e. $z_T \geq 0$. □

Theorem 9.13

Let T be a closed densely defined operator on $\mathcal{H}$. Then there exists a unique pair of operators (u, K) such that

(1) $T = uK$,

(2) *K is positive and self-adjoint,*

(3) *$u^* u$ is the projection onto the closure of the range of K.*

Proof

Let $z_T = u|z_T|$ be the polar decomposition of the bounded operator z_T. Since $\big\| |z_T| \big\| = \|z_T\| \leq 1$ and

$$\ker \left(\mathbb{1} - |z_T|^* |z_T| \right) = \ker(\mathbb{1} - z_T{}^* z_T),$$

the operator $|z_T|$ is the z-transform of some closed densely defined operator K: $|z_T| = z_K$. As $|z_T|$ is positive, we have $K = K^*$ and K is positive. Also $\mathsf{D}(K) = \mathsf{D}(T)$, because

$$\mathsf{D}(K) = (\mathbb{1} - z_K{}^* z_K)^{\frac{1}{2}}\mathcal{H} = (\mathbb{1} - |z_T|^*|z_T|)^{\frac{1}{2}}\mathcal{H} = (\mathbb{1} - z_T{}^* z_T)^{\frac{1}{2}}\mathcal{H} = \mathsf{D}(T).$$

Moreover

$$\begin{aligned}\mathsf{G}(K) &= \left\{\begin{bmatrix}(\mathbb{1} - z_K{}^* z_K)^{\frac{1}{2}}\xi \\ z_K\xi\end{bmatrix}\,\middle|\, \xi \in \mathcal{H}\right\} \\ &= \left\{\begin{bmatrix}(\mathbb{1} - z_T{}^* z_T)^{\frac{1}{2}}\xi \\ |z_T|\xi\end{bmatrix}\,\middle|\, \xi \in \mathcal{H}\right\}\end{aligned} \tag{9.9}$$

and obviously

$$\mathsf{G}(T) = \left\{\begin{bmatrix}(\mathbb{1} - z_T{}^* z_T)^{\frac{1}{2}}\xi \\ z_T\xi\end{bmatrix}\,\middle|\, \xi \in \mathcal{H}\right\}.$$

It follows that

$$\mathsf{G}(T) = \begin{bmatrix}\mathbb{1} & 0 \\ 0 & u\end{bmatrix}\mathsf{G}(K),$$

i.e. $T = uK$.

Finally u^*u is the projection onto the closure of the range of $|z_T|$ and formula (9.9) shows that the latter coincides with the range of K.

Let us now address the uniqueness of the pair (u, K). Let (v, D) be another pair such that

- $T = vD$,
- D is positive and self-adjoint,
- v^*v is the projection onto the closure of the range of D.

Then, on one hand, using Proposition 8.14 we obtain

$$T^*T = (uK)^*(uK) = Ku^*uK = K^2,$$

and on the other hand

$$T^*T = (vD)^*(vD) = Dv^*vD = D^2.$$

It follows that

$$z_T = T(\mathbb{1} + T^*T)^{-\frac{1}{2}} = T(\mathbb{1} + K^2)^{-\frac{1}{2}} = T(\mathbb{1} + D^2)^{-\frac{1}{2}}$$

and consequently

$$uK(\mathbb{1} + K^2)^{-\frac{1}{2}} = vD(\mathbb{1} + D^2)^{-\frac{1}{2}}$$

or, in other words,

$$uz_K = vz_D.$$

Applying the operation $x \mapsto x^*x$ to both sides of this equality yields

$$z_K u^* u z_K = z_D v^* v z_D.$$

Since the range of the z-transform coincides with the range of the operator, we have $u^*uz_K = z_K$ and $v^*vz_D = z_D$ and therefore

$$z_K{}^2 = z_D{}^2.$$

The operators z_K and z_D are positive, so by uniqueness of positive square roots we get $z_K = z_D$ and thus $K = D$.

The equality $u = v$ follows now from $uK = vK$ and the fact that $u^*u = v^*v$ is the projection onto the closure of the range of K in the same way as for bounded operators. □

The decomposition $T = uK$ of a closed densely defined operator T obtained in Theorem 9.13 is called the *polar decomposition* of T. The partial isometry u is sometimes referred to as the *phase* of T, while the positive self-adjoint operator K is called the *modulus* of *absolute value* of T and is denoted by the symbol $|T|$.

Notes

The concept of the z-transform was introduced for the first time in a framework far more general than the theory of operators on Hilbert spaces. Given any C*-algebra A consider a vector space $\mathcal{E}$ endowed with a right action of A and a "scalar product" with values in A. The latter is a sesquilinear map $\langle\cdot|\cdot\rangle : \mathcal{E} \times \mathcal{E} \to \mathsf{A}$ possessing several properties which guarantee that $\langle\cdot|\cdot\rangle$ defines a norm on $\mathcal{E}$ in a way analogous to how it is defined on spaces with a genuine scalar product:

$$\|\xi\|_{\mathcal{E}} = \sqrt{\|\langle\xi|\xi\rangle\|}, \qquad \xi \in \mathcal{E}$$

(for details see [Lan, Chapter 1]). If $\mathcal{E}$ is complete in this norm then it is called a Hilbert C*-module over A. In particular, Hilbert C*-modules over $\mathsf{A} = \mathbb{C}$ are nothing else than Hilbert spaces. It turns out that many problems of pure mathematics require the additional generality of Hilbert C*-modules over non-trivial C*-algebras instead of Hilbert spaces.

A surprising feature of Hilbert C*-modules is that not all linear operators on such a module necessarily have an adjoint, even bounded ones. The situation becomes even more complicated when we analyze unbounded operators. The notion of z-transform turned out to be crucial for the development of the theory of operators on Hilbert C*-modules. This theory is described very thoroughly in the book [Lan]. Many interesting analogies and differences between studying operators on Hilbert spaces and on Hilbert C*-modules are highlighted in the paper [WoNa].

Spectral Theorems

P. Sołtan, *A Primer on Hilbert Space Operators*, Compact Textbooks in Mathematics,
https://doi.org/10.1007/978-3-319-92061-0_10

Just as in the case of bounded operators, spectral theorem for unbounded operators assumes various forms. We will begin with continuous functional calculus which can be defined exclusively using the z-transform. Next we will move on to Borel functional calculus and finally assign to each self-adjoint operator its spectral measure and discuss functional calculus for unbounded functions.

10.1 Continuous Functional Calculus

Let T be a self-adjoint operator on $\mathcal{H}$. We will begin by defining functional calculus for *bounded* continuous functions. The space $\mathrm{C_b}(\mathbb{R})$ of bounded continuous functions on $\mathbb{R}$ is a C*-algebra with natural operations of addition and multiplications, the uniform norm $\|\cdot\|_\infty$ and complex conjugation of functions as involution. Functional calculus should be a unital $*$-homomorphism $\mathrm{C_b}(\mathbb{R}) \to \mathsf{B}(\mathcal{H})$. Moreover, putting

$$\boldsymbol{\zeta}(t) = \tfrac{t}{\sqrt{1+t^2}}, \qquad t \in \mathbb{R}$$

we define a continuous and bounded function on $\mathbb{R}$ (in fact $\boldsymbol{\zeta}$ is a homeomorphism $\mathbb{R} \to]-1, 1[$) and we would like to have $\boldsymbol{\zeta}(T) = z_T$.

Since z_T is a self-adjoint (bounded) operator, we can consider the map

$$\mathrm{C_b}(\mathbb{R}) \ni f \longmapsto (f \circ \boldsymbol{\zeta}^{-1})(z_T)$$

which formally would satisfy our requirements. However, the function $f \circ \boldsymbol{\zeta}^{-1}$ belongs to $\mathrm{C_b}(]-1, 1[)$ while $\sigma(z_T) \subset [-1, 1]$[1] and hence the construction of continuous functional calculus for T will have to be more involved.

[1]Moreover, one can show that if $\sigma(z_T) \subset]-1, 1[$ then T is bounded.

Theorem 10.1
Let T be a self-adjoint operator. Then there exists a unique unital $*$-homomorphism $\mathrm{C}_b(\mathbb{R}) \to \mathrm{B}(\mathcal{H})$ denoted by

$$\mathrm{C}_b(\mathbb{R}) \ni f \longmapsto f(T) \in \mathrm{B}(\mathcal{H})$$

such that $\boldsymbol{\zeta}(T) = z_T$.

Proof
For any $f \in \mathrm{C}_b(\mathbb{R})$ the function

$$]-1, 1[\ni t \longmapsto f\big(\boldsymbol{\zeta}^{-1}(t)\big)(1-t^2)^{\frac{1}{2}}$$

extends to a continuous function on $[-1, 1]$. Let us denote this extension by $\widetilde{f}$. Now take $\xi \in \mathrm{D}(T)$. Then there exists a unique $\eta \in \mathcal{H}$ such that $\xi = (\mathbb{1} + T^*T)^{-\frac{1}{2}}\eta = (\mathbb{1} - z_T{}^* z_T)^{\frac{1}{2}}\eta$ (see Eq. (9.6)). Consider the map

$$\mathrm{D}(T) \ni \xi \longmapsto \widetilde{f}(z_T)\eta \in \mathcal{H}. \tag{10.1}$$

We will show that it is bounded. Indeed: the function $\widetilde{f}$ satisfies

$$\big(\overline{\widetilde{f}}\widetilde{f}\big)(t) = \big|\widetilde{f}(t)\big|^2 \le \|f\|_\infty^2(1-t^2), \qquad t \in [-1, 1].$$

Thus, using properties of continuous functional calculus for bounded self-adjoint operators, we obtain

$$\widetilde{f}(z_T)^*\widetilde{f}(z_T) \le \|f\|_\infty^2(\mathbb{1} - z_T{}^* z_T).$$

Therefore

$$\begin{aligned}\|\widetilde{f}(z_T)\eta\|^2 &= \big\langle \widetilde{f}(z_T)\eta \big| \widetilde{f}(z_T)\eta \big\rangle = \big\langle \eta \big| \widetilde{f}(z_T)^*\widetilde{f}(z_T)\eta \big\rangle \\ &\le \|f\|_\infty^2 \big\langle \eta \big| (\mathbb{1} - z_T{}^* z_T)\eta \big\rangle \\ &= \|f\|_\infty^2 \Big\langle (\mathbb{1} - z_T{}^* z_T)^{\frac{1}{2}}\eta \Big| (\mathbb{1} - z_T{}^* z_T)^{\frac{1}{2}}\eta \Big\rangle = \|f\|_\infty^2 \|\xi\|^2.\end{aligned}$$

It follows that the map (10.1) extends to a bounded operator $\mathcal{H} \to \mathcal{H}$ with norm not larger than $\|f\|_\infty$. Let us denote it by the symbol $f(T)$. Clearly the map $f \mapsto f(T)$ is linear and if f is constant and equal to 1, then $f(T) = \mathbb{1}$. We will now show that $f(T)g(T) = (fg)(T)$. To that end note that for any $h \in \mathrm{C}_b(\mathbb{R})$ we have (by definition)

$$h(T)(\mathbb{1} - z_T{}^* z_T)^{\frac{1}{2}}\phi = \widetilde{h}(z_T)\phi, \qquad \phi \in \mathcal{H}, \tag{10.2}$$

so that for any $\xi \in \mathsf{D}(T^*T) = (\mathbb{1} - z_T{}^* z_T)\mathcal{H}$

$$\begin{aligned} f(T)g(T)\xi &= f(T)g(z_T)(\mathbb{1} - z_T{}^* z_T)\phi \\ &= f(T)g(T)(\mathbb{1} - z_T{}^* z_T)^{\frac{1}{2}}(\mathbb{1} - z_T{}^* z_T)^{\frac{1}{2}}\phi \\ &= f(T)\widetilde{g}(z_T)(\mathbb{1} - z_T{}^* z_T)^{\frac{1}{2}}\phi \\ &= f(T)(\mathbb{1} - z_T{}^* z_T)^{\frac{1}{2}}\widetilde{g}(z_T)\phi = \widetilde{f}(z_T)\widetilde{g}(z_T)\phi, \end{aligned}$$

where ϕ is the vector satisfying $\xi = (\mathbb{1} - z_T{}^* z_T)\phi$. Furthermore, for any $t \in \,]-1, 1[$

$$\begin{aligned} \widetilde{f}(t)\widetilde{g}(t) &= f\big(\zeta^{-1}(t)\big)(1 - t^2)^{\frac{1}{2}} g\big(\zeta^{-1}(t)\big)(1 - t^2)^{\frac{1}{2}} \\ &= (fg)\big(\zeta^{-1}(t)\big)(1 - t^2)^{\frac{1}{2}}(1 - t^2)^{\frac{1}{2}} = (\widetilde{fg})(t)(1 - t^2)^{\frac{1}{2}}, \end{aligned}$$

and so

$$\begin{aligned} f(T)g(T)\xi &= (\widetilde{f}\widetilde{g})(z_T)(\mathbb{1} - z_T{}^* z_T)^{\frac{1}{2}}\phi \\ &= (fg)(T)(\mathbb{1} - z_T{}^* z_T)^{\frac{1}{2}}(\mathbb{1} - z_T{}^* z_T)^{\frac{1}{2}}\phi \\ &= (fg)(T)(\mathbb{1} - z_T{}^* z_T)\phi = (fg)(T)\xi. \end{aligned}$$

This proves that $f(T)g(T) = (fg)(T)$, because $\mathsf{D}(T^*T)$ is dense in $\mathcal{H}$. Finally the equality $\overline{f}(T) = f(T)^*$ follows directly from the definition of the operator $f(T)$.

We have shown that $f \mapsto f(T)$ is a unital $*$-homomorphism from $\mathrm{C_b}(\mathbb{R})$ to $\mathsf{B}(\mathcal{H})$. Of course, for $f = \zeta$ we have $\widetilde{f}(t) = t(1 - t^2)^{\frac{1}{2}}$, so that for $\xi = (\mathbb{1} - z_T{}^* z_T)^{\frac{1}{2}}\eta \in \mathsf{D}(T)$ we get

$$f(T)\xi = \widetilde{f}(z_T)\eta = z_T(\mathbb{1} - z_T{}^* z_T)^{\frac{1}{2}}\phi = z_T\xi$$

which means that $f(T) = z_T$.

Let us now address the uniqueness of a homomorphism satisfying the condition of the theorem. Let $\Phi : \mathrm{C_b}(\mathbb{R}) \to \mathsf{B}(\mathcal{H})$ be a unital $*$-homomorphism such that $\Phi(\zeta) = z_T$. Then the values of Φ are uniquely determined on polynomials in ζ. Moreover, we know from Proposition 4.9 (▶ Sect. 4.2) that any unital $*$-homomorphism from a Banach $*$-algebra to $\mathsf{B}(\mathcal{H})$ is continuous, so that the values of Φ are also uniquely determined on all functions of the form $\varphi \circ \zeta$ with $\varphi \in \mathrm{C}([-1, 1])$. In other words, the condition that $\Phi(\zeta) = z_T$ uniquely determines values of Φ on all functions from $\mathrm{C_b}(\mathbb{R})$ possessing limits at $\pm\infty$.

We will see that this already determines the homomorphism Φ uniquely on all of $\mathrm{C_b}(\mathbb{R})$. For $f \in \mathrm{C_b}(\mathbb{R})$ define

$$f_n(t) = \begin{cases} f(-n), & t < -n, \\ f(t), & -n \leq t \leq n, \\ f(n), & t > n, \end{cases} \qquad t \in \mathbb{R}.$$

We will prove that

$$\Phi(f_n)\xi \xrightarrow[n\to\infty]{} \Phi(f)\xi, \qquad \xi \in \mathsf{D}(T) \tag{10.3}$$

which guarantees that the operator $\Phi(f)$ is uniquely determined on a dense subset of $\mathcal{H}$ by values of Φ on functions which have limits at $\pm\infty$. This, in turn, gives us uniqueness of continuous functional calculus[2]

Let $g(t) = \left(1 - \zeta(t)^2\right)^{\frac{1}{2}}$. Since values of Φ on polynomials in z_T are uniquely determined, we obtain

$$\Phi(g^2) = \mathbb{1} - z_T{}^* z_T$$

and thus, by uniqueness of positive square roots, we get

$$\Phi(g) = (\mathbb{1} - z_T{}^* z_T)^{\frac{1}{2}}.$$

Now for $\xi \in \mathsf{D}(T)$ there exists a unique $\eta \in \mathcal{H}$ such that $\xi = (\mathbb{1} - z_T{}^* z_T)^{\frac{1}{2}}\eta$, i.e.

$$\xi = \Phi(g)\eta$$

and it follows that

$$\Phi(f_n)\xi = \Phi(f_n)\Phi(g)\eta = \Phi(f_n g)\eta.$$

Finally we have $f_n g \xrightarrow[n\to\infty]{} fg$ uniformly on $\mathbb{R}$, because

$$\left|(f_n g)(t) - (fg)(t)\right| \leq 2\|f\|_\infty \sup_{|t|>n} g(t) \xrightarrow[n\to\infty]{} 0,$$

and consequently $\Phi(f_n g) \xrightarrow[n\to\infty]{} \Phi(fg)$ in $\mathsf{B}(\mathcal{H})$. Therefore

$$\Phi(f_n)\xi = \Phi(f_n g)\eta \xrightarrow[n\to\infty]{} \Phi(fg)\eta = \Phi(f)\xi$$

and (10.3) is proved. □

[2]Clearly, since $\left\|\Phi(f_n)\right\| < C$ for some constant C, we also have

$$\Phi(f_n)\xi \xrightarrow[n\to\infty]{} \Phi(f)\xi, \qquad \xi \in \mathcal{H}.$$

Indeed: fix $\xi \in \mathcal{H}$ and for $\varepsilon > 0$ choose $\xi' \in \mathsf{D}(T)$ such that $\|\xi - \xi'\| < \frac{\varepsilon}{3C}$. Then

$$\begin{aligned}\left\|\Phi(f_n)\xi - \Phi(f)\xi\right\| &\leq \left\|\Phi(f_n)\xi - \Phi(f_n)\xi'\right\| + \left\|\Phi(f_n)\xi' - \Phi(f)\xi'\right\| + \left\|\Phi(f)\xi' - \Phi(f)\xi\right\| \\ &\leq C\|\xi - \xi'\| + \left\|\Phi(f_n)\xi' - \Phi(f)\xi'\right\| + C\|\xi - \xi'\| < \tfrac{2\varepsilon}{3M} + \left\|\Phi(f_n)\xi' - \Phi(f)\xi'\right\|.\end{aligned}$$

Therefore, for n large enough, so that $\left\|\Phi(f_n)\xi' - \Phi(f)\xi'\right\| < \frac{\varepsilon}{3C}$, the above estimate gives $\left\|\Phi(f_n)\xi - \Phi(f)\xi\right\| < \varepsilon$.

Lemma 10.2 *Let* $f \in \mathrm{C}_b(\mathbb{R})$ *be real-valued and such that* $f(t) \neq 0$ *for all* $t \in \mathbb{R}$. *Then* $\ker f(T) = \{0\}$.

Proof
Denote

$$a = f(T), \qquad b = \widetilde{f}(z_T), \qquad c = (\mathbb{1} - {z_T}^* z_T)^{\frac{1}{2}}.$$

Let us first see that $\ker b = \{0\}$. For this we can assume that z_T is an operator of multiplication by a function on a Hilbert space $L_2(\Omega, \mu)$, i.e. $z_T = M_g$. Furthermore, since $\ker(\mathbb{1} - {z_T}^2) = \ker(\mathbb{1} - {z_T}^* z_T) = \{0\}$, we have $\mu(\{\omega \,|\, g(\omega) = \pm 1\}) = 0$ (Proposition 4.4). The operator $b = \widetilde{f}(z_T)$ is then the operator of multiplication by $\widetilde{f} \circ g$. As $\widetilde{f}(t) \neq 0$ for $t \in]-1, 1[$, we get

$$\mu(\{\omega \,|\, (\widetilde{f} \circ g)(\omega) = 0\}) = 0$$

and it follows that $\ker b = \{0\}$ (again by Proposition 4.4).

By definition of functional calculus for T we have

$$ac\eta = b\eta, \qquad \eta \in \mathcal{H},$$

i.e. $ac = b$. The operators a, b and c are self-adjoint, so

$$ca = (ac)^* = b^* = b = ac.$$

Thus $b = ac = ca$ and consequently $\ker a \subset \ker b = \{0\}$. □

Lemma 10.2 makes it simple to extend functional calculus of bounded continuous functions to all continuous functions on $\mathbb{R}$. For simplicity we will restrict attention to real-valued functions. This will also make it easier to make a connection with Borel functional calculus which will be presented in the next section.

Theorem 10.3
Let T *be a self-adjoint operator. Then for any real-valued* $f \in \mathrm{C}(\mathbb{R})$ *there exists a unique closed densely defined operator* $f(T)$ *such that* $z_{f(T)} = (\boldsymbol{\zeta} \circ f)(T)$. *Moreover* $f(T)$ *is self-adjoint.*

Proof
Let $z = (\boldsymbol{\zeta} \circ f)(T)$. We have $\|z\| \leq \|\boldsymbol{\zeta}\|_\infty = 1$ and $\ker(\mathbb{1} - z^* z) = \{0\}$ by Lemma 10.2, as

$$\mathbb{1} - z^* z = g(T),$$

where $g(t) = 1 - \zeta\big(f(t)\big)^2 > 0$ for all $t \in \mathbb{R}$. It follows that z is a z-transform of a unique closed densely defined operator which we call $f(T)$. This operator is self-adjoint, because $z = z^*$.

□

10.2 Borel Functional Calculus

Let T be a self-adjoint operator on $\mathcal{H}$. In view of Theorem 9.10, the z-transform z_T of T is a self-adjoint operator with spectrum contained in $[-1, 1]$. In particular z_T is unitarily equivalent to a multiplication operator (Theorem 4.7). This allows us to conclude the following:

Theorem 10.4
Let T be a self-adjoint operator on $\mathcal{H}$. Then there exist a semifinite measure space (Ω, μ), a measurable real-valued function f on Ω and a unitary operator $u \in \mathsf{B}\big(L_2(\Omega, \mu), \mathcal{H}\big)$ such that $T = uM_f u^*$, i.e.
(1) $\mathsf{D}(T) = \big\{u\psi \,\big|\, \psi \in L_2(\Omega, \mu),\ f\psi \in L_2(\Omega, \mu)\big\}$,
(2) for $u\psi \in \mathsf{D}(T)$ we have $Tu\psi = uf\psi$.
Moreover $\sigma(T) = V_{ess}(f)$.

Proof
Since z_T is a self-adjoint operator, by Theorem 4.7 there exits a semifinite measure space (Ω, μ), a measurable real-valued function F on Ω and a unitary operator $u \in \mathsf{B}\big(L_2(\Omega, \mu), \mathcal{H}\big)$ such that $z_T = uM_F u^*$. Consider the operator $S = u^*Tu$. Then by Remark 9.11 we have $z_S = u^* z_T u = M_F$. It follows that

$$\begin{aligned}\mathsf{D}(S) &= \big\{(\mathbb{1} - z_S{}^* z_S)^{\frac{1}{2}}\phi \,\big|\, \phi \in L_2(\Omega, \mu)\big\}\\ &= \big\{(\mathbb{1} - M_F{}^2)^{\frac{1}{2}}\phi \,\big|\, \phi \in L_2(\Omega, \mu)\big\}\\ &= \big\{M_{\sqrt{1-F^2}}\phi \,\big|\, \phi \in L_2(\Omega, \mu)\big\}.\end{aligned}$$

Note that since $\ker(\mathbb{1} - z_S{}^* z_S) = \{0\}$, by Proposition 4.4 we have $\sqrt{1 - F^2} \neq 0$ almost everywhere on Ω. It follows that the function $f = \frac{F}{\sqrt{1-F^2}}$ is finite almost everywhere. Moreover, for $\psi = \sqrt{1 - F^2}\phi \in \mathsf{D}(S)$ we have $S\psi = F\phi = f\psi$, so

$$S \subset M_f.$$

Both S and M_f are self-adjoint, so by Remark 8.9 they are equal. It follows that $T = uM_f u^*$.

Let us analyze the spectrum of T. We know that $\sigma(T) = \sigma(u^*Tu) = \sigma(M_f)$, so the theorem will be proved when we show that $\sigma(M_f) = V_{ess}(f)$. Let $\lambda \in \mathbb{C} \setminus V_{ess}(f)$. Then there exits $r > 0$ such that

$$\{\omega \,|\, |\lambda - f(\omega)| < r\}$$

is of measure zero. Therefore the function $g = \frac{1}{\lambda - f}$ belongs to $L_\infty(\Omega, \mu)$ and satisfies $\|g\|_\infty \leq \frac{1}{r}$. In particular the operator M_g is bounded. Furthermore

- for $\varphi \in \mathsf{D}(M_f)$ we have $(\lambda \mathbb{1} - M_f)\varphi = (\lambda - f)\varphi$ and this immediately shows that $M_g(\lambda \mathbb{1} - M_f)\varphi = \varphi$,
- for $\psi \in L_2(\Omega, \mu)$ we have $M_g \psi \in \mathsf{D}(M_f)$, because

$$|fg| = \left|\frac{f}{\lambda - f}\right| = \left|\frac{f - \lambda + \lambda}{\lambda - f}\right| = \left|\lambda \frac{1}{\lambda - f} - 1\right| \leq |\lambda||g| + 1,$$

and so $|fg\psi| \leq \big(|\lambda||g| + 1\big)|\psi|$, i.e. $fg\psi \in L_2(\Omega, \mu)$.

It follows that $M_g = (\lambda \mathbb{1} - M_f)^{-1}$.

Conversely, if $\lambda \in V_{\text{ess}}(f)$ then for any $n \in \mathbb{N}$ the set

$$\{\omega \,|\, |\lambda - f(\omega)| \leq \tfrac{1}{n}\}$$

is of strictly positive measure and consequently it contains a subset Λ_n of strictly positive and finite measure. Putting $\psi_n = \frac{1}{\sqrt{\mu(\Lambda_n)}} \chi_{\Lambda_n}$ we obtain a sequence of norm 1 vectors such that $\psi_n \in \mathsf{D}(M_f)$ for all n, as $|f| \leq |\lambda| + \frac{1}{n}$ on Λ_n, and thus

$$\int_\Omega |f\psi_n|^2 \, d\mu = \frac{1}{\mu(\Lambda_n)} \int_{\Lambda_n} |f|^2 \, d\mu \leq \big(|\lambda| + \tfrac{1}{n}\big)^2 < +\infty.$$

It is easy to see that $\big\|(\lambda \mathbb{1} - M_f)\psi_n\big\|_2 \leq \frac{1}{n}$ and hence there does not exist a bounded operator a such that $a(\lambda \mathbb{1} - M_f) = \mathbb{1}$, i.e. $\lambda \in \sigma(M_f)$. □

As Theorem 10.4 says, any self-adjoint operator T is unitarily equivalent to an operator of multiplication by a real-valued measurable function. In particular $\sigma(T)$ is a non-empty subset of $\mathbb{R}$.

As in the case of bounded operators, Theorem 10.4 allows us to extend functional calculus for a self-adjoint operator from bounded continuous functions to bounded Borel functions on $\mathbb{R}$.

Theorem 10.5

Let T be a self-adjoint operator on $\mathcal{H}$ and let $\mathscr{B}(\mathbb{R})$ denote the set of bounded Borel functions on $\mathbb{R}$. Then there exists a unique unital $*$-homomorphism $\mathscr{B}(\mathbb{R}) \to \mathsf{B}(\mathcal{H})$ denoted by

$$\mathscr{B}(\mathbb{R}) \ni g \longmapsto g(T) \in \mathsf{B}(\mathcal{H})$$

(Continued)

Theorem 10.5 (continued)
such that
- $\zeta(T) = z_T$,
- *if* $(g_n)_{n\in\mathbb{N}}$ *is a uniformly bounded sequence of Borel functions converging pointwise to* g *then* $g_n(T) \xrightarrow[n\to\infty]{} g(T)$ *in strong topology.*

Proof
Just as in the proof of the analogous statement for bounded operators, we use the fact that T is unitarily equivalent to an operator of multiplication by a real-valued measurable function on a Hilbert space $L_2(\Omega, \mu)$ for some semifinite measure space (Ω, μ). In fact one can assume that Ω is a locally compact topological space, μ is a Borel measure and f is Borel. Then for any bounded Borel function g on $\mathbb{R}$ the function $g \circ f$ is Borel and bounded on Ω. For such g we define $g(T)$ as $uM_{g\circ f}u^*$.

It follows immediately from this definition that $g \mapsto g(T)$ is a contractive unital $*$-homomorphism and that

$$\|g(T)\| = \|M_{g\circ f}\| = \sup\left\{|\lambda| \,\middle|\, \lambda \in V_{\text{ess}}(g \circ f)\right\} \le \|g\|_\infty.$$

Furthermore, for $g = \zeta$ we have $g(T) = u^*M_{\frac{f}{\sqrt{1+f^2}}}u$, i.e. $g(T) = z_T$ (cf. proof of Theorem 10.4 and Remark 9.11). Finally, if $(g_n)_{n\in\mathbb{N}}$ is a bounded sequence of Borel functions converging pointwise to g and $\xi \in \mathcal{H}$ then

$$\begin{aligned}\left\|g_n(x)\xi - g(x)\xi\right\|^2 &= \|uM_{(g_n-g)\circ f}u^*\xi\|^2 \\ &= \left\|M_{(g_n-g)\circ f}u^*\xi\right\|^2 \\ &= \int_\Omega \left|g_n(f(\omega)) - g(f(\omega))\right|^2 \left|(u^*\xi)(\omega)\right|^2 d\mu(\omega) \xrightarrow[n\to\infty]{} 0.\end{aligned}$$

by the dominated convergence theorem.

The proof of uniqueness of the homomorphism

$$\mathscr{B}(\sigma(x)) \ni f \longmapsto f(x) \in \mathsf{B}(\mathcal{H})$$

follows the lines of the proof of the corresponding statement for bounded operators (Theorem 4.10). We know that a unital $*$-homomorphism $\mathrm{C_b}(\mathbb{R}) \to \mathsf{B}(\mathcal{H})$ mapping ζ to z_T is unique, so any unital $*$-homomorphism $\Phi : \mathscr{B}(\mathbb{R}) \to \mathsf{B}(\mathcal{H})$ satisfying $\Phi(\zeta) = z_T$ must coincide with the continuous functional calculus for T on $\mathrm{C_b}(\mathbb{R})$. Using this, just as in the proof of Theorem 4.10, we show that the family

$$\mathcal{L} = \left\{\Delta \in \mathfrak{M} \,\middle|\, \Phi(\chi_\Delta) = \chi_\Delta(x)\right\}$$

(1) contains all open sets (in particular $\Omega \in \mathcal{L}$),
(2) is closed with respect to taking complements,
(3) is closed with respect to countable unions of pairwise disjoint elements,

i.e. $\mathcal{L}$ is a $\boldsymbol{\lambda}$-system (see Appendix A.2). Since the family of all open subsets of $\mathbb{R}$ is a $\boldsymbol{\pi}$-system, by Dynkin's theorem on $\boldsymbol{\pi}$- and $\boldsymbol{\lambda}$-systems (Theorem A.2), the $\boldsymbol{\sigma}$-algebra of all Borel subsets of $\mathbb{R}$ is contained in $\mathcal{L}$. Now any bounded Borel function is a pointwise limit of a bounded sequence of simple Borel functions, so the homomorphisms Φ and $f \mapsto f(x)$ must agree on all of $\mathscr{B}(\mathbb{R})$. □

10.3 Spectral Measure

Let T be a self-adjoint operator on $\mathcal{H}$. Our aim in this section is to express T as an integral of the function $\lambda \mapsto \lambda$ with respect to certain spectral measure on $\mathbb{R}$. So far we have discussed such integrals only for bounded functions (▶ Sect. 4.3) and now we will deal with arbitrary Borel functions $\mathbb{R} \to \mathbb{C}$. We will freely use concepts and notation introduced in ▶ Sect. 4.3.

Let E be a spectral measure on a measurable space $(\Omega, \mathfrak{M})$.

Lemma 10.6 *For any $\varphi, \psi \in \mathcal{H}$ and bounded measurable function g on Ω we have*

$$\left| \int_\Omega g \, d\langle \varphi \mid E\psi \rangle \right| \leq \|\varphi\| \left(\int_\Omega |g|^2 \, d\langle \psi \mid E\psi \rangle \right)^{\frac{1}{2}}.$$

Proof
Let $|\langle \varphi | E\psi \rangle|$ be the variation of the measure $\langle \varphi | E\psi \rangle$.[3] It follows from the Radon-Nikodym theorem that there exists a measurable function u with values of modulus 1 such that

$$ug \; \langle \varphi | E\psi \rangle = |g| \, |\langle \varphi | E\psi \rangle|.$$

[3]In ▶ Sect. 4.2 we recalled the notion of total variation of a measure. Now the *variation* of a complex measure ν defined on a σ-algebra $\mathfrak{N}$ is the map $|\nu| : \mathfrak{N} \to [0 + \infty]$ defined by

$$|\nu|(\Delta) = \sup \left\{ \sum_{n=1}^{N} |\nu(\Delta_n)| \right\}, \qquad \Delta \in \mathfrak{N}$$

with the supremum is taken over all measurable partitions of Δ. Then $|\nu|$ is a measure and ν is absolutely continuous with respect to $|\nu|$. Also

$$\left| \int f \, d\nu \right| \leq \int |f| \, d|\nu|$$

for any integrable f.

Therefore

$$\left|\int_\Omega g\, d\langle\varphi\,|\,E\psi\rangle\right| \le \int_\Omega |g|\, d|\langle\varphi|E\psi\rangle|$$
$$= \int_\Omega ug\, d\langle\varphi\,|\,E\psi\rangle = \langle\varphi|x_{ug}\psi\rangle = |\langle\varphi|x_{ug}\psi\rangle| \le \|\varphi\|\|x_{ug}\psi\|,$$

and since

$$\|x_{ug}\psi\|^2 = \langle\psi|x_{|ug|^2}\psi\rangle = \langle\psi|x_{|g|^2}\psi\rangle = \int_\Omega |g|^2\, d\langle\psi\,|\,E\psi\rangle,$$

we find that

$$\left|\int_\Omega g\, d\langle\varphi\,|\,E\psi\rangle\right| \le \|\varphi\|\left(\int_\Omega |g|^2\, d\langle\psi\,|\,E\psi\rangle\right)^{\frac{1}{2}}.$$

□

Before going on to define integrals of not necessarily bounded functions with respect to spectral measures let us note that for *bounded* measurable functions f and g we have

$$\int_\Omega gf\, d\langle\varphi\,|\,E\psi\rangle = \langle\varphi|x_f x_g\psi\rangle = \langle x_{\overline{f}}\varphi|x_g\psi\rangle = \int_\Omega g\, d\langle x_{\overline{f}}\varphi\,|\,E\psi\rangle,$$

which implies the following equality of complex measures:

$$\langle x_{\overline{f}}\varphi|E\psi\rangle = f\,\langle\varphi|E\psi\rangle\,. \tag{10.4}$$

Proposition 10.7 *Let f be a measurable function $\Omega \to \mathbb{C}$ and define*

$$\mathscr{D}_f = \left\{\psi \in \mathcal{H}\,\middle|\, \int_\Omega |f|^2\, d\langle\psi\,|\,E\psi\rangle < +\infty\right\}.$$

Then $\mathscr{D}_f$ is a dense subspace of $\mathcal{H}$.

Proof

For $n \in \mathbb{N}$ let $\Lambda_n = \{\omega \in \Omega \,|\, |f(\omega)| \le n\}$. For $\psi \in E(\Lambda_n)\mathcal{H}$ and any measurable $\Delta \subset \Omega$ we have

$$E(\Delta)\psi = E(\Delta)E(\Lambda_n)\psi = E(\Delta \cap \Lambda_n)\psi,$$

so that

$$\langle\psi|E\psi\rangle(\Delta) = \langle\psi|E(\Delta)\psi\rangle = \langle\psi|E(\Delta\cap\Lambda_n)\psi\rangle = \langle\psi|E\psi\rangle(\Delta\cap\Lambda_n).$$

Hence

$$\int_\Omega |f|^2\, d\langle\psi \mid E\psi\rangle = \int_{\Lambda_n} |f|^2\, d\langle\psi \mid E\psi\rangle \le n^2\|\psi\|^2 < +\infty$$

and thus $E(\Lambda_n)\mathcal{H} \subset \mathscr{D}_f$ for all $n \in \mathbb{N}$. On the other hand, since $\chi_{\Lambda_n} \xrightarrow[n\to\infty]{} 1$ pointwise, we have $E(\Lambda_n)\psi \xrightarrow[n\to\infty]{} \psi$. It follows that $\mathscr{D}_f$ is a dense subset of $\mathcal{H}$.

In order to see that $\mathscr{D}_f$ is a vector subspace, take $\psi, \varphi \in \mathscr{D}_f$. Then for any $\Delta \in \mathfrak{M}$

$$\|E(\Delta)(\psi+\varphi)\|^2 \le \left(\|E(\Delta)\psi\| + \|E(\Delta)\varphi\|\right)^2 \le 2\|E(\Delta)\psi\|^2 + 2\|E(\Delta)\varphi\|^2.$$

In other words

$$\langle\psi+\varphi|E(\psi+\varphi)\rangle \le 2\,\langle\psi|E\psi\rangle + 2\,\langle\varphi|E\varphi\rangle$$

which shows that $\psi+\varphi \in \mathscr{D}_f$. The fact that $\mathscr{D}_f$ is closed under scalar multiplication is clear. □

For a function f as in Proposition 10.7 we will now define an operator x_f with domain $\mathsf{D}(x_f) = \mathscr{D}_f$. Let us take $\psi \in \mathsf{D}(x_f)$ and $\varphi \in \mathcal{H}$. By Lemma 10.6 and dominated convergence theorem, we have

$$\left|\int_\Omega f\, d\langle\varphi \mid E\psi\rangle\right| \le \|\varphi\|\left(\int_\Omega |f|^2\, d\langle\psi \mid E\psi\rangle\right)^{\frac{1}{2}},$$

and consequently

$$\varphi \longmapsto \int_\Omega f\, d\langle\varphi \mid E\psi\rangle$$

is a bounded liner functional on $\mathcal{H}$. It follows that there exists a unique vector η such that

$$\int_\Omega f\, d\langle\varphi \mid E\psi\rangle = \langle\varphi|\eta\rangle\,.$$

We put $x_f\psi = \eta$. The left hand side of the above expression is linear in ψ, so x_f is a densely defined linear operator. Moreover, for f bounded, this definition of x_f coincides with the definition of x_f given in Theorem 10.5 (see also ▶ Sect. 4.3).

Remark 10.8 Let f_1 and f_2 be measurable functions $\Omega \to \mathbb{C}$ and let $\psi \in \mathsf{D}(x_{f_1}) \cap \mathsf{D}(x_{f_2}) = \mathscr{D}_{f_1} \cap \mathscr{D}_{f_2}$. Then for any $\varphi \in \mathcal{H}$

$$\begin{aligned}\langle\varphi|x_{f_1}\psi\rangle + \langle\varphi|x_{f_2}\psi\rangle &= \int_\Omega f_1\, d\langle\varphi \mid E\psi\rangle + \int_\Omega f_2\, d\langle\varphi \mid E\psi\rangle \\ &= \int_\Omega (f_1+f_2)\, d\langle\varphi \mid E\psi\rangle = \langle\varphi|x_{f_1+f_2}\psi\rangle,\end{aligned}$$

i.e. we have $x_{f_1}\psi + x_{f_2}\psi = x_{f_1+f_2}\psi$.

Theorem 10.9
Let E be a spectral measure on $(\Omega, \mathfrak{M})$. *Then*
(1) *for any measurable function* $f : \Omega \to \mathbb{C}$ *the operator* x_f *is closed,*
(2) *for* $\psi \in \mathsf{D}(x_f)$ *we have*

$$\|x_f\psi\|^2 = \int_\Omega |f|^2\, d\langle\psi \,|\, E\psi\rangle, \tag{10.5}$$

(3) *for any measurable* $f, g : \Omega \to \mathbb{C}$ *we have* $x_f x_g \subset x_{fg}$ *and, moreover,* $\mathsf{D}(x_f x_g) = \mathscr{D}_g \cap \mathscr{D}_{fg}$,
(4) *for any measurable* $f : \Omega \to \mathbb{C}$ *we have* $x_f{}^* = x_{\overline{f}}$.

Proof
We will prove (2), (3) and (4). Statement (1) follows by applying (4) to $\overline{f}$.

Let χ_n be the characteristic function of $\{\omega \in \Omega \,\big|\, |f(\omega)| \leq n\}$ and let $f_n = \chi_n f$. As f_n is bounded, we have $\mathscr{D}_{f-f_n} = \mathscr{D}_f$. Furthermore, by the dominated convergence theorem, for $\psi \in \mathscr{D}_f$ we have

$$\left\|x_f\psi - x_{f_n}\psi\right\|^2 \leq \int_\Omega |f - f_n|\, d\langle\psi \,|\, E\psi\rangle \xrightarrow[n\to\infty]{} 0 \tag{10.6}$$

(cf. Remark 10.8 and the paragraph preceding it). Since f_n is bounded, we can perform the following computation

$$\left\|x_{f_n}\psi\right\|^2 = \langle\psi \big| x_{|f_n|^2}\psi\rangle = \int_\Omega |f_n|^2\, d\langle\psi \,|\, E\psi\rangle,$$

and consequently (10.5) follows from the approximation (10.6) and (2) is proved.

Assume for the moment that f is bounded. Then $\mathscr{D}_g \subset \mathscr{D}_{fg}$ and for $\psi \in \mathscr{D}_g$, $\varphi \in \mathcal{H}$ we have

$$\langle\varphi \big| x_f x_g\psi\rangle = \left\langle x_{\overline{f}}\varphi \big| x_g\psi\right\rangle = \int_\Omega g\, d\langle x_{\overline{f}}\varphi \,|\, E\psi\rangle = \int_\Omega fg\, d\langle\varphi \,|\, E\psi\rangle = \langle\varphi \big| x_{fg}\psi\rangle$$

by (10.4). This shows that for $\psi \in \mathscr{D}_g$ (and f bounded) we have

$$x_f x_g\psi = x_{fg}\psi,$$

and consequently

$$\int_\Omega |f|^2 \, d\langle x_g\psi \mid E x_g\psi\rangle = \|x_f x_g \psi\|^2 = \|x_{fg}\psi\|^2 = \int_\Omega |fg|^2 \, d\langle \psi \mid E\psi\rangle, \qquad \psi \in \mathsf{D}(g(T)).$$

Now since the above formula holds for all bounded measurable functions, it also holds for any measurable function $f : \Omega \to \mathbb{C}$. Therefore $x_g\psi$ belongs to $\mathsf{D}(x_f) = \mathscr{D}_f$ if and only if $\psi \in \mathscr{D}_{fg}$, i.e.

$$\mathsf{D}(x_f x_g) = \mathscr{D}_g \cap \mathscr{D}_{fg}.$$

This proves Statement (3).

For the proof of (4) we will once more use the approximation of f by the bounded functions f_n. Take $\psi \in \mathscr{D}_f$ and $\varphi \in \mathscr{D}_{\overline{f}} = \mathscr{D}_f$. Then

$$\langle \varphi | x_f \psi \rangle = \lim_{n\to\infty} \langle \varphi | x_{f_n} \psi \rangle = \lim_{n\to\infty} \left\langle x_{\overline{f_n}} \varphi \middle| \psi \right\rangle = \left\langle x_{\overline{f}} \varphi \middle| \psi \right\rangle$$

which shows that $\varphi \in \mathsf{D}(x_f{}^*)$ and $x_f{}^*\psi = x_{\overline{f}}\psi$. In other words

$$x_{\overline{f}} \subset x_f{}^*. \tag{10.7}$$

Now take $\zeta \in \mathsf{D}(x_f{}^*)$. By (3) we have $x_{f_n} = x_f x_{\chi_n}$, so that

$$x_{\chi_n} x_f{}^* \subset (x_f x_{\chi_n})^* = x_{f_n}{}^* = x_{\overline{f_n}},$$

as x_{χ_n} is a self-adjoint operator. In particular

$$x_{\chi_n} x_f{}^* \zeta = x_{\overline{f_n}} \zeta, \qquad n \in \mathbb{N}.$$

We have

$$\int_\Omega |f_n|^2 \, d\langle \zeta \mid E\zeta\rangle = \|x_{\overline{f_n}}\zeta\|^2 = \|x_{\chi_n} x_f{}^*\zeta\|^2 = \int_\Omega |\chi_n|^2 \, d\langle x_f{}^*\zeta \mid E x_f{}^*\zeta\rangle \le \left\langle x_f{}^*\zeta \middle| E x_f{}^*\zeta \right\rangle(\Omega) < +\infty$$

and hence

$$\int_\Omega |f|^2 \, d\langle \zeta \mid E\zeta\rangle < +\infty.$$

It follows that $\zeta \in \mathscr{D}_{\overline{f}} = \mathsf{D}(x_{\overline{f}})$. This way we proved that $\mathsf{D}(x_f{}^*) \subset \mathsf{D}(x_{\overline{f}})$ which together with (10.7) gives (4).

□

Theorem 10.10

Let T be a self-adjoint operator. Then there exists a unique spectral measure E_T on $\mathbb{R}$ such that

$$T = \int_{\mathbb{R}} \lambda \, dE_T(\lambda).$$

Before proving Theorem 10.10 let us note that the result allows us to define functional calculus for T for all (not necessarily bounded) Borel functions on $\mathbb{R}$. More precisely for a Borel function $f : \mathbb{R} \to \mathbb{C}$ we put

$$f(T) = \int_{\mathbb{R}} f(\lambda) \, dE(\lambda). \tag{10.8}$$

In case f is bounded, formula (10.8) produces the image of f under functional calculus for bounded Borel functions defined in Theorem 10.5. Moreover, it follows easily form Theorem 10.9(3) that if f and g are Borel functions $\mathbb{R} \to \mathbb{C}$ and g is bounded then $f(T)g(T) = (fg)(T)$. Indeed: we have

$$\mathsf{D}\big(f(T)g(T)\big) = \mathscr{D}_g \cap \mathscr{D}_{fg} = \mathcal{H} \cap \mathscr{D}_{fg} = \mathsf{D}\big((fg)(T)\big),$$

and therefore

$$g(T)f(T) \subset (gf)(T) = (fg)(T) = f(T)g(T). \tag{10.9}$$

One cannot expect equality in (10.9), because e.g. if $f(T)$ is unbounded and $g = 0$, then the domain of the left hand side is $\mathsf{D}\big(f(T)\big)$, while the right hand side is defined on all of $\mathcal{H}$.

Proof of Theorem 10.10

Let E_{z_T} be the spectral measure of the z-transform of T. Since $\sigma(z_T) \subset [-1, 1]$ we can treat E_{z_T} as a Borel spectral measure on $[-1, 1]$. Moreover, we immediately see that since ± 1 is not an eigenvalue of z_T, we have $E_{z_T}\big(\{\pm 1\}\big) = 0$. In other words E_{z_T} is a spectral measure on $]-1, 1[$.

Let

$$S = \int_{-1}^{1} \frac{\mu}{\sqrt{1-\mu^2}} \, dE_{z_T}(\mu).$$

Let us check that $S = T$. Using the notation introduced above, we have $S = x_f$, where $f(\mu) = \frac{\mu}{\sqrt{1-\mu^2}}$. Put $g(\mu) = \sqrt{1-\mu^2}$. Then

$$Sx_g \subset x_{fg} = \int_{-1}^{1} \mu \, dE_{z_T}(\mu) = z_T$$

by Theorem 10.9(3). Moreover $\mathsf{D}(Sx_g) = \mathscr{D}_g \cap \mathscr{D}_{fg} = \mathcal{H}$, as g and fg are bounded on $\Omega =]-1, 1[$. This, in turn, means that

$$Sx_g = z_T. \tag{10.10}$$

Finally note that

$$x_g = g(z_T) = (\mathbb{1} - z_T{}^2)^{\frac{1}{2}},$$

which shows that (10.10) is in fact

$$S(\mathbb{1} - z_T{}^2)^{\frac{1}{2}} = T(\mathbb{1} - z_T{}^2)^{\frac{1}{2}}.$$

In particular $T \subset S$, and since S and T are self-adjoint, we obtain $S = T$ by Remark 8.9.

Now let E_T be the push-forward of E_{z_T} by the map $\boldsymbol{\zeta}^{-1} :]-1, 1[\to \mathbb{R}$, i.e.

$$E_T(\Delta) = E_{z_T}\big(\boldsymbol{\zeta}(\Delta)\big), \qquad \Delta \in \mathfrak{M},$$

where $\mathfrak{M}$ is the $\boldsymbol{\sigma}$-algebra of Borel subsets of $\mathbb{R}$ (in standard notation we write $E_T = (\boldsymbol{\zeta}^{-1})_* E_{z_T}$). Then

$$T = \int_{-1}^{1} \frac{\mu}{\sqrt{1-\mu^2}} \, dE_{z_T}(\mu) = \int_{-1}^{1} \boldsymbol{\zeta}^{-1}(\mu) \, dE_{z_T}(\mu) = \int_{\mathbb{R}} \lambda \, dE_T(\lambda).$$

We move on to the question of uniqueness of the spectral measure E_T. Suppose that E is a Borel spectral measure on $\mathbb{R}$ such that

$$T = \int_{\mathbb{R}} \lambda \, dE(\lambda).$$

Then $z_T = \boldsymbol{\zeta}(T) = \int_{\mathbb{R}} \frac{\lambda}{\sqrt{1+\lambda^2}} \, dE(\lambda)$. Now let E' be the push-forward of E onto $]-1, 1[$ by the map $\boldsymbol{\zeta}$. Then

$$z_T = \int_{-1}^{1} \mu \, dE'(\mu).$$

It follows from the uniqueness of the spectral measure of z_T that $E' = E_{z_T}$. Since ζ is a homeomorphism, we obtain

$$E = (\zeta^{-1})_* E' = (\zeta^{-1})_* E_{z_T},$$

so that $E = E_T$. □

Let T be a self-adjoint operator and let f be a continuous real-valued function on $\mathbb{R}$. In ▶ Sect. 10.1. we defined $f(T)$ as the operator whose z-transform is $(\zeta \circ f)(T)$ (having defined bounded continuous functions of T in Theorem 10.1). The obvious question is whether this operator is the same as $f(T)$ defined by formula (10.8). The following reasoning yields a positive answer to this question: let E_T be the spectral measure of T and take a real-valued $f \in C(\mathbb{R})$. Since

$$\int_{\mathbb{R}} \lambda \, d(f_* E_T)(\lambda) = \int_{\mathbb{R}} f(\lambda) \, dE_T(\lambda) = f(T),$$

it follows from uniqueness of the spectral measure of $f(T)$ that $E_{f(T)} = f_* E_T$. Therefore

$$z_{f(T)} = \int_{\mathbb{R}} \zeta(\mu) \, dE_{f(T)}(\mu) = \int_{\mathbb{R}} (\zeta \circ f)(\lambda) \, dE_T(\lambda) = (\zeta \circ f)(T)$$

where in the second equality we used a well-known property of a push-forward of measure.

Consequently the z-transform of the operator $f(T)$ (defined by formula (10.8)) is the operator declared as the z-transform of $f(T)$ in Theorem 10.3.

Notes

Spectral theory of unbounded operators has numerous applications and the reader will find far reaching extensions of the basic material of this chapter e.g. in [AkGl, Chapter VI], [Mau, Chapter VI], [ReSi$_1$, Chapter VIII] and in selected chapters of [Ped, Rud$_2$]. All of these textbooks and monographs contain many examples and exercises on the subject.

Self-Adjoint Extensions of Symmetric Operators

P. Sołtan, *A Primer on Hilbert Space Operators*, Compact Textbooks in Mathematics,
https://doi.org/10.1007/978-3-319-92061-0_11

Self-adjointness of certain operators is crucial in many applications (e.g. in quantum physics or in partial differential equations). However, operators which naturally occur in many problems turn out to be symmetric, but not necessarily self-adjoint. It is for that reason that the problem of existence and classification of self-adjoint extensions of symmetric operators was one of the first challenges of the theory of unbounded operators on Hilbert spaces. In this chapter we will present the first results on that topic.

11.1 Containment of Operators in Terms of z-Transforms

Theorem 11.1
Let S and T be densely defined closed operators on $\mathcal{H}$. Then $T \subset S$ if and only if

$$(\mathbb{1} - z_S z_S{}^*)^{\frac{1}{2}} z_T = z_S(\mathbb{1} - z_T{}^* z_T)^{\frac{1}{2}}.$$

Proof
The relation $T \subset S$ means exactly that $\mathsf{G}(T) \subset \mathsf{G}(S)$. Now the graphs of S and T have a precise description in terms of their z-transforms given in Theorem 9.6:

$$\mathsf{G}(T) = \left\{ \begin{bmatrix} (\mathbb{1} - z_T{}^* z_T)^{\frac{1}{2}}\xi \\ z_T\xi \end{bmatrix} \middle| \, \xi \in \mathcal{H} \right\}, \quad \mathsf{G}(S) = \left\{ \begin{bmatrix} (\mathbb{1} - z_S{}^* z_S)^{\frac{1}{2}}\xi \\ z_S\xi \end{bmatrix} \middle| \, \xi \in \mathcal{H} \right\}.$$

Moreover

$$\mathsf{G}(T) = U_{z_T}\big(\mathcal{H} \oplus \{0\}\big), \quad \mathsf{G}(S) = U_{z_S}\big(\mathcal{H} \oplus \{0\}\big),$$

where

$$U_{z_T} = \begin{bmatrix} (\mathbb{1} - z_T{}^* z_T)^{\frac{1}{2}} & -z_T{}^* \\ z_T & (\mathbb{1} - z_T z_T{}^*)^{\frac{1}{2}} \end{bmatrix},$$

$$U_{z_S} = \begin{bmatrix} (\mathbb{1} - z_S{}^* z_S)^{\frac{1}{2}} & -z_S{}^* \\ z_S & (\mathbb{1} - z_S z_S{}^*)^{\frac{1}{2}} \end{bmatrix}$$

are unitary operators on $\mathcal{H} \oplus \mathcal{H}$ and $\mathcal{H} \oplus \{0\}$ denotes the subspace of horizontal vectors

$$\left\{ \begin{bmatrix} \xi \\ 0 \end{bmatrix} \,\middle|\, \xi \in \mathcal{H} \right\} \subset \mathcal{H} \oplus \mathcal{H}.$$

(cf. remarks following Theorem 9.9). It follows that $\mathsf{G}(T) \subset \mathsf{G}(S)$ is equivalent to

$$U_{z_T}\big(\mathcal{H} \oplus \{0\}\big) \subset U_{z_S}\big(\mathcal{H} \oplus \{0\}\big). \tag{11.1}$$

Applying $U_{z_S}{}^*$ to both sides of (11.1) we obtain

$$U_{z_S}{}^* U_{z_T}\big(\mathcal{H} \oplus \{0\}\big) \subset \mathcal{H} \oplus \{0\}.$$

It is easy to see that the only operators on $\mathcal{H} \oplus \mathcal{H}$ preserving the subspace $\mathcal{H} \oplus \{0\}$ are the ones which in matrix notation have the form

$$\begin{bmatrix} \bullet & \bullet \\ 0 & \bullet \end{bmatrix}.$$

In particular, the condition $T \subset S$ is equivalent to vanishing of the lower left corner of the matrix

$$\begin{aligned} U_{z_S}{}^* U_{z_T} &= \begin{bmatrix} (\mathbb{1} - z_S{}^* z_S)^{\frac{1}{2}} & -z_S{}^* \\ z_S & (\mathbb{1} - z_S z_S{}^*)^{\frac{1}{2}} \end{bmatrix}^* \begin{bmatrix} (\mathbb{1} - z_T{}^* z_T)^{\frac{1}{2}} & -z_T{}^* \\ z_T & (\mathbb{1} - z_T z_T{}^*)^{\frac{1}{2}} \end{bmatrix} \\ &= \begin{bmatrix} (\mathbb{1} - z_S{}^* z_S)^{\frac{1}{2}} & z_S{}^* \\ -z_S & (\mathbb{1} - z_S z_S{}^*)^{\frac{1}{2}} \end{bmatrix} \begin{bmatrix} (\mathbb{1} - z_T{}^* z_T)^{\frac{1}{2}} & -z_T{}^* \\ z_T & (\mathbb{1} - z_T z_T{}^*)^{\frac{1}{2}} \end{bmatrix} \\ &= \begin{bmatrix} \bullet & \bullet \\ -z_S(\mathbb{1} - z_T{}^* z_T)^{\frac{1}{2}} + (\mathbb{1} - z_S z_S{}^*)^{\frac{1}{2}} z_T & \bullet \end{bmatrix} \end{aligned}$$

In other words $T \subset S$ if and only if $(\mathbb{1} - z_S z_S{}^*)^{\frac{1}{2}} z_T = z_S(\mathbb{1} - z_T{}^* z_T)^{\frac{1}{2}}$. □

11.2 Cayley Transform

Before we continue, let us introduce a piece of notation which adds certain flexibility in dealing with partial isometries. Let $v \in \mathsf{B}(\mathcal{H})$ be a partial isometry. We know form Proposition 3.12 that there is a closed subspace $\mathcal{S} \subset \mathcal{H}$ such that $v\xi = 0$ for $\xi \in \mathcal{S}^{\perp}$ and $\|v\xi\| = \|\xi\|$ for $\xi \in \mathcal{S}$. We will denote by $\mathring{v}$ the operator $v\big|_{\mathcal{S}}$ understood as an operator on $\mathcal{H}$ with domain $\mathsf{D}(\mathring{v}) = \mathcal{S} = v^*\mathcal{H}$. In particular we can speak of the graph of $\mathring{v}$ as a subset of $\mathcal{H} \oplus \mathcal{H}$. In accordance with our previous conventions for $x \in \mathsf{B}(\mathcal{H})$ the operator $\mathring{v} + x$ is defined on $\mathsf{D}(\mathring{v} + x) = \mathsf{D}(\mathring{v}) = \mathcal{S}$. The operation $v \mapsto \mathring{v}$ is a bijection between the set of partial isometries on $\mathcal{H}$ and the set of isometric operators defined on closed subspaces of $\mathcal{H}$.

From now on let T be a symmetric operator. We are interested in *self-adjoint extensions* of T, i.e. operators S such that $T \subset S$ and $S = S^*$. Since self-adjoint operators are closed and symmetric ones are closable (see Proposition 8.7), any self-adjoint extension of T is an extension of $\overline{T}$. We can therefore assume that T is closed.

It follows from Theorem 11.1 that

$$(\mathbb{1} - {z_T}^* z_T)^{\frac{1}{2}} z_T = {z_T}^* (\mathbb{1} - {z_T}^* z_T)^{\frac{1}{2}}. \tag{11.2}$$

Define two operators

$$w_{\pm} = z_T \pm \mathrm{i}(\mathbb{1} - {z_T}^* z_T)^{\frac{1}{2}}.$$

Then w_+ and w_- are isometries:

$$\begin{aligned} {w_{\pm}}^* w_{\pm} &= \big({z_T}^* \mp \mathrm{i}(\mathbb{1} - {z_T}^* z_T)^{\frac{1}{2}}\big)\big(z_T \pm \mathrm{i}(\mathbb{1} - {z_T}^* z_T)^{\frac{1}{2}}\big) \\ &= {z_T}^* z_T \mp \mathrm{i}\big((\mathbb{1} - {z_T}^* z_T)^{\frac{1}{2}} z_T - {z_T}^*(\mathbb{1} - {z_T}^* z_T)^{\frac{1}{2}}\big) + (\mathbb{1} - {z_T}^* z_T) = \mathbb{1} \end{aligned}$$

by (11.2). Let $\mathscr{W}_{\pm} = w_{\pm}\mathcal{H}$ and $\mathscr{D}_{\pm} = \mathscr{W}_{\pm}^{\perp}$.

The subspaces $\mathscr{D}_+$ and $\mathscr{D}_-$ are called the *deficiency subspaces* of T and their dimensions $n_{\pm} = \dim \mathscr{D}_{\pm}$ are the *deficiency indices* of T.

Proposition 11.2 *We have* $\mathscr{D}_{\pm} = \ker(T^* \mp \mathrm{i}\mathbb{1})$.

Proof
Since $\mathscr{D}_{\pm} = \mathscr{W}_{\pm}^{\perp}$, the condition $\zeta \in \mathscr{D}_{\pm}$ means that

$$\left\langle \zeta \,\middle|\, z_T\xi \pm \mathrm{i}(\mathbb{1} - {z_T}^* z_T)^{\frac{1}{2}}\xi \right\rangle = 0, \qquad \xi \in \mathcal{H}. \tag{11.3}$$

Taking into account the description of the graph of T in terms of z_T given in Theorem 9.6, we see that (11.3) is equivalent to

$$\langle \zeta \,|\, T\psi \pm \mathrm{i}\psi \rangle = 0, \qquad \psi \in \mathsf{D}(T),$$

which means precisely that $\zeta \in \mathsf{D}\big((T \pm \mathrm{i}\mathbb{1})^*\big)$ and $(T \pm \mathrm{i}\mathbb{1})^*\psi = 0$. By Proposition 8.15(2) this amounts to $\zeta \in \ker(T^* \mp \mathrm{i}\mathbb{1})$. □

Let $c_T = w_- w_+{}^*$. Then c_T is a partial isometry with initial subspace $\mathscr{W}_+$ and final subspace $\mathscr{W}_-$. The operator $\mathring{c_T}$ is called the *Cayley transform* of the closed symmetric operator T.

Using once again the argument from the proof of Proposition 11.2 we easily obtain

$$\begin{aligned}\mathsf{G}(\mathring{c_T}) &= \left\{\begin{bmatrix}\vartheta \\ w_- w_+{}^*\vartheta\end{bmatrix} \,\middle|\, \vartheta \in \mathscr{W}_+\right\} \\ &= \left\{\begin{bmatrix}w_+\xi \\ w_-\xi\end{bmatrix} \,\middle|\, \xi \in \mathcal{H}\right\} \\ &= \left\{\begin{bmatrix}T\psi + \mathrm{i}\psi \\ T\psi - \mathrm{i}\psi\end{bmatrix} \,\middle|\, \psi \in \mathsf{D}(T)\right\} = \begin{bmatrix}\mathrm{i}\mathbb{1} & \mathbb{1} \\ -\mathrm{i}\mathbb{1} & \mathbb{1}\end{bmatrix}\mathsf{G}(T)\end{aligned}$$

and hence

$$\mathsf{G}(T) = \begin{bmatrix}-\mathrm{i}\mathbb{1} & \mathrm{i}\mathbb{1} \\ \mathbb{1} & \mathbb{1}\end{bmatrix}\mathsf{G}(\mathring{c_T}).$$

In particular $\mathring{c_T}$ determines T.

As $\mathsf{G}(T)$ is a graph of a densely defined operator, by Propositions 8.3 and 8.1

$$\mathsf{G}(T)^\perp \cap \big(\mathcal{H} \oplus \{0\}\big) = \mathsf{G}(T) \cap \big(\{0\} \oplus \mathcal{H}\big) = \{0\}.$$

Therefore, for any $\xi \in \mathcal{H}$ we have

$$\left(\begin{bmatrix}\xi \\ 0\end{bmatrix} \perp \begin{bmatrix}-\mathrm{i}\mathbb{1} & \mathrm{i}\mathbb{1} \\ \mathbb{1} & \mathbb{1}\end{bmatrix}\mathsf{G}(\mathring{c_T})\right) \Longrightarrow \big(\xi = 0\big). \tag{11.4}$$

The condition on the left hand side of (11.4) means

$$\left\langle \begin{bmatrix}\xi \\ 0\end{bmatrix} \,\middle|\, \begin{bmatrix}-\mathrm{i}\mathbb{1} & \mathrm{i}\mathbb{1} \\ \mathbb{1} & \mathbb{1}\end{bmatrix}\begin{bmatrix}\vartheta \\ \mathring{c_T}\vartheta\end{bmatrix}\right\rangle = 0, \qquad \vartheta \in \mathsf{D}(\mathring{c_T})$$

which can be rewritten as

$$\big\langle \xi \,\big|\, (\mathring{c_T} - \mathbb{1})\vartheta \big\rangle = 0, \qquad \vartheta \in \mathscr{W}_+.$$

In other words, a vector which is orthogonal to $(\mathring{c_T} - \mathbb{1})\mathscr{W}_+ = (c_T - \mathbb{1})c_T{}^*\mathcal{H}$ must be zero, i.e. the subspace $(c_T - \mathbb{1})c_T{}^*\mathcal{H}$ is dense in $\mathcal{H}$:

$$\overline{(c_T - \mathbb{1})c_T{}^*\mathcal{H}} = \mathcal{H}. \tag{11.5}$$

Notice further, that it follows from (11.5) that $\ker(c_T - \mathbb{1}) = \{0\}$. Indeed: noting that for any partial isometry v we have $\ker\big(v^*(v - \mathbb{1})\big) = \ker\big(v(v^* - \mathbb{1})\big)$, we obtain $\ker(c_T - \mathbb{1}) \subset \ker\big(c_T{}^*(c_T - \mathbb{1})\big) = \ker\big(c_T(c_T{}^* - \mathbb{1})\big) = \big((\mathbb{1} - c_T)c_T{}^*\mathcal{H}\big)^{\perp} = \{0\}$.[1]

Theorem 11.3
The mapping $T \mapsto c_T$ is a bijection between the set of closed symmetric operators on $\mathcal{H}$ and the set of partial isometries $c \in \mathsf{B}(\mathcal{H})$ satisfying $\overline{(c - \mathbb{1})c^*\mathcal{H}} = \mathcal{H}$.

Proof
We have already shown that if T is a closed symmetric operator then c_T is a partial isometry satisfying (11.5) and that the map $T \mapsto c_T$ is injective (in other words c_T determines T). It remains to prove that if c is a partial isometry such that $\overline{(c - \mathbb{1})c^*\mathcal{H}} = \mathcal{H}$ then there exists a closed symmetric operator T such that $c = c_T$.

Let

$$\mathsf{G} = \begin{bmatrix} -i\mathbb{1} & i\mathbb{1} \\ \mathbb{1} & \mathbb{1} \end{bmatrix} \mathsf{G}(\mathring{c}).$$

It is clear that G is a closed subspace of $\mathcal{H} \oplus \mathcal{H}$. By reversing the reasoning leading up to condition (11.5), we obtain

$$\left(\begin{bmatrix} \xi \\ 0 \end{bmatrix} \perp \mathsf{G} \right) \Longrightarrow \left(\xi = 0 \right).$$

More precisely, if

$$\left\langle \begin{bmatrix} \xi \\ 0 \end{bmatrix} \middle| \begin{bmatrix} -i\mathbb{1} & i\mathbb{1} \\ \mathbb{1} & \mathbb{1} \end{bmatrix} \begin{bmatrix} \vartheta \\ \mathring{c}\vartheta \end{bmatrix} \right\rangle = 0, \qquad \vartheta \in \mathsf{D}(\mathring{c}),$$

then

$$\xi \perp (c - \mathbb{1})\vartheta, \qquad \vartheta \in c^*\mathcal{H},$$

i.e. $\xi \perp (c - \mathbb{1})c^*\mathcal{H}$ and this implies $\xi = 0$. Similarly we check that also the condition

$$\left(\begin{bmatrix} 0 \\ \eta \end{bmatrix} \in \mathsf{G} \right) \Longrightarrow \left(\eta = 0 \right)$$

[1] Triviality of the kernel of $c_T - \mathbb{1}$ follows also from the fact that $\mathsf{G}(T)$ does not contain non-zero vertical vectors, in a way similar to how we derived (11.5) from (11.4). It is not surprising that one of these conditions implies the other, because T is symmetric, i.e.

$$\mathsf{G}(T) \subset \mathsf{G}(T^*) = \left[\begin{smallmatrix} 0 & \mathbb{1} \\ -\mathbb{1} & 0 \end{smallmatrix}\right] \mathsf{G}(T)^{\perp}.$$

holds. Indeed: if

$$\begin{bmatrix} 0 \\ \eta \end{bmatrix} = \begin{bmatrix} -i\mathbb{1} & i\mathbb{1} \\ \mathbb{1} & \mathbb{1} \end{bmatrix} \begin{bmatrix} \vartheta \\ \mathring{c}\vartheta \end{bmatrix}$$

for some $\vartheta \in \mathsf{D}(\mathring{c})$ then $\eta = (\mathbb{1} + \mathring{c})\vartheta$ and $(\mathring{c} - \mathbb{1})\vartheta = 0$. Thus

$$\vartheta \in \ker(\mathring{c} - \mathbb{1}) \subset \ker(c - \mathbb{1}) \subset \ker\big(c^*(c - \mathbb{1})\big) = \ker\big(c(c^* - \mathbb{1})\big) = \big((c - \mathbb{1})c^*\mathcal{H}\big)^\perp = \{0\}$$

(see remarks preceding the statement of the theorem) and consequently $\eta = (\mathbb{1} + \mathring{c})\vartheta = 0$.

This way we have shown that G is a graph of a closed densely defined operator T (Corollary 8.4).

Let us now look a the graph of T^*:

$$\begin{aligned} \mathsf{G}(T^*) &= \begin{bmatrix} 0 & \mathbb{1} \\ -\mathbb{1} & 0 \end{bmatrix} \mathsf{G}(T)^\perp = \begin{bmatrix} 0 & \mathbb{1} \\ -\mathbb{1} & 0 \end{bmatrix} \left(\begin{bmatrix} -i\mathbb{1} & i\mathbb{1} \\ \mathbb{1} & \mathbb{1} \end{bmatrix} \mathsf{G}(\mathring{c}) \right)^\perp \\ &= \left(\begin{bmatrix} 0 & \mathbb{1} \\ -\mathbb{1} & 0 \end{bmatrix} \begin{bmatrix} -i\mathbb{1} & i\mathbb{1} \\ \mathbb{1} & \mathbb{1} \end{bmatrix} \mathsf{G}(\mathring{c}) \right)^\perp = \left(\begin{bmatrix} \mathbb{1} & \mathbb{1} \\ i\mathbb{1} & -i\mathbb{1} \end{bmatrix} \mathsf{G}(\mathring{c}) \right)^\perp. \end{aligned}$$

Now take an arbitrary $\left[\begin{smallmatrix} \xi \\ \eta \end{smallmatrix}\right] \in \mathsf{G}(T)$. Then there exists $\vartheta \in \mathsf{D}(\mathring{c})$ such that $\left[\begin{smallmatrix} \xi \\ \eta \end{smallmatrix}\right] = \left[\begin{smallmatrix} -i\mathbb{1} & i\mathbb{1} \\ \mathbb{1} & \mathbb{1} \end{smallmatrix}\right]\left[\begin{smallmatrix} \vartheta \\ \mathring{c}\vartheta \end{smallmatrix}\right]$ and for any $\vartheta' \in \mathsf{D}(\mathring{c})$ we have

$$\begin{aligned} &\left\langle \begin{bmatrix} -i\mathbb{1} & i\mathbb{1} \\ \mathbb{1} & \mathbb{1} \end{bmatrix} \begin{bmatrix} \vartheta \\ \mathring{c}\vartheta \end{bmatrix} \middle| \begin{bmatrix} \mathbb{1} & \mathbb{1} \\ i\mathbb{1} & -i\mathbb{1} \end{bmatrix} \begin{bmatrix} \vartheta' \\ \mathring{c}\vartheta' \end{bmatrix} \right\rangle \\ &\qquad = \left\langle \begin{bmatrix} \vartheta \\ \mathring{c}\vartheta \end{bmatrix} \middle| \begin{bmatrix} i\mathbb{1} & \mathbb{1} \\ -i\mathbb{1} & \mathbb{1} \end{bmatrix} \begin{bmatrix} \mathbb{1} & \mathbb{1} \\ i\mathbb{1} & -i\mathbb{1} \end{bmatrix} \begin{bmatrix} \vartheta' \\ \mathring{c}\vartheta' \end{bmatrix} \right\rangle \\ &\qquad = \left\langle \begin{bmatrix} \vartheta \\ \mathring{c}\vartheta \end{bmatrix} \middle| \begin{bmatrix} 2i\mathbb{1} & 0 \\ 0 & -2i\mathbb{1} \end{bmatrix} \begin{bmatrix} \vartheta' \\ \mathring{c}\vartheta' \end{bmatrix} \right\rangle \\ &\qquad = 2i\big(\langle \vartheta | \vartheta' \rangle - \langle \mathring{c}\vartheta | \mathring{c}\vartheta' \rangle\big) = 0, \end{aligned}$$

as $\mathring{c}$ preserves the scalar product on $\mathsf{D}(\mathring{c})$. This proves that

$$\mathsf{G}(T) \subset \mathsf{G}(T^*),$$

i.e. T is symmetric. □

Remark 11.4 Let c_1 and c_2 be partial isometries on $\mathcal{H}$ such that $\mathring{c_1} \subset \mathring{c_2}$. Then if $\overline{(c_1 - \mathbb{1})c_1{}^*\mathcal{H}} = \mathcal{H}$ then also $\overline{(c_2 - \mathbb{1})c_2{}^*\mathcal{H}} = \mathcal{H}$. Indeed: we have $c_2{}^*\mathcal{H} = \mathsf{D}(\mathring{c_2}) \supset \mathsf{D}(\mathring{c_1}) = c_1{}^*\mathcal{H}$ and for any $\xi \in \mathsf{D}(\mathring{c_1})$ we have $(c_2 - \mathbb{1})\xi = (c_1 - \mathbb{1})\xi$. Therefore $(c_2 - \mathbb{1})c_2{}^*\mathcal{H} \supset (c_1 - \mathbb{1})c_1{}^*\mathcal{H}$.

Corollary 11.5 *Fix a closed symmetric operator T. Then T' is a closed symmetric extension of T if and only if $\mathring{c}_T \subset \mathring{c}_{T'}$.*

Proof
We already know that partial isometries c satisfying

$$\overline{(c-\mathbb{1})c^*\mathcal{H}} = \mathcal{H}$$

are in bijection with closed symmetric operators and the operator corresponding to c is T_c defined by

$$\mathsf{G}(T_c) = \begin{bmatrix} -\mathrm{i}\mathbb{1} & \mathrm{i}\mathbb{1} \\ \mathbb{1} & \mathbb{1} \end{bmatrix} \mathsf{G}(\mathring{c}). \tag{11.6}$$

Let s be a partial isometry such that $\mathring{c} \subset \mathring{s}$. Since $\overline{(c-\mathbb{1})c^*\mathcal{H}} = \mathcal{H}$, Remark 11.4 shows that also $\overline{(s-\mathbb{1})s^*\mathcal{H}} = \mathcal{H}$. Furthermore, it is clear from formula (11.6) that the operator T_s defined by s satisfies $\mathsf{G}(T_c) \subset \mathsf{G}(T_s)$ or, in other words, $T_c \subset T_s$. □

Theorem 11.6
Let T be a closed symmetric operator with deficiency subspaces $\mathscr{D}_+$ and $\mathscr{D}_-$. Let

$$\widetilde{\mathscr{D}}_\pm = \left\{ \begin{bmatrix} \phi \\ \pm\mathrm{i}\phi \end{bmatrix} \,\middle|\, \phi \in \mathscr{D}_\pm \right\}.$$

Then

$$\mathsf{G}(T^*) = \mathsf{G}(T) \oplus \widetilde{\mathscr{D}}_+ \oplus \widetilde{\mathscr{D}}_-. \tag{11.7}$$

Proof
Let us first check that the three subspaces on the right hand side of (11.7) are pairwise orthogonal.

It is clear that for any $\xi, \eta \in \mathcal{H}$ we have

$$\left\langle \begin{bmatrix} \xi \\ \mathrm{i}\xi \end{bmatrix} \,\middle|\, \begin{bmatrix} \eta \\ -\mathrm{i}\eta \end{bmatrix} \right\rangle = \langle \xi \,|\, \eta \rangle + \langle \mathrm{i}\xi \,|\, -\mathrm{i}\eta \rangle = 0,$$

and hence the subspaces $\widetilde{\mathscr{D}}_+$ and $\widetilde{\mathscr{D}}_-$ are orthogonal.

Take $\psi \in \mathsf{D}(T)$ and $\phi \in \mathscr{D}_\pm = \ker(T^* \mp \mathrm{i}\mathbb{1})$. Then

$$\left\langle \begin{bmatrix} \psi \\ T\psi \end{bmatrix} \middle| \begin{bmatrix} \phi \\ \pm\mathrm{i}\phi \end{bmatrix} \right\rangle = \langle \psi | \phi \rangle \pm \mathrm{i} \langle T\psi | \phi \rangle$$

$$= \langle \psi | \phi \rangle \pm \mathrm{i} \langle \psi | T^*\phi \rangle = \langle \psi | \phi \rangle \pm \mathrm{i} \langle \psi | \pm \mathrm{i}\phi \rangle = 0$$

which shows that $\mathsf{G}(T)$ is orthogonal to $\widetilde{\mathscr{D}}_+$ and to $\widetilde{\mathscr{D}}_-$.

As T is symmetric, we have $\mathsf{G}(T) \subset \mathsf{G}(T^*)$. Moreover $\widetilde{\mathscr{D}}_\pm \subset \mathsf{G}(T^*)$ by definition of these subspaces ($\mathscr{D}_\pm$ are eigenspaces of T^*). It remains to show that if a vector $\left[\begin{smallmatrix} \varphi \\ T^*\varphi \end{smallmatrix} \right] \in \mathsf{G}(T^*)$ is orthogonal to $\mathsf{G}(T) \oplus \widetilde{\mathscr{D}}_+ \oplus \widetilde{\mathscr{D}}_-$ then $\varphi = 0$. Let us therefore take $\varphi \in \mathsf{D}(T^*)$ such that

$$\begin{bmatrix} \varphi \\ T^*\varphi \end{bmatrix} \perp \mathsf{G}(T) \oplus \widetilde{\mathscr{D}}_+ \oplus \widetilde{\mathscr{D}}_-.$$

Then for any $\psi \in \mathsf{D}(T)$ we have

$$0 = \left\langle \begin{bmatrix} \varphi \\ T^*\varphi \end{bmatrix} \middle| \begin{bmatrix} \psi \\ T\psi \end{bmatrix} \right\rangle = \langle \varphi | \psi \rangle + \langle T^*\varphi | T\psi \rangle,$$

and so the functional

$$\mathsf{D}(T) \ni \psi \longmapsto \langle T^*\varphi | T\psi \rangle = - \langle \varphi | \psi \rangle$$

is continuous. It follows that $T^*\varphi \in \mathsf{D}(T^*)$ and $T^{*2}\varphi = -\varphi$, and hence

$$\varphi \in \mathsf{D}\big((T^* + \mathrm{i}\mathbb{1})(T^* - \mathrm{i}\mathbb{1})\big)$$

and

$$(T^* + \mathrm{i}\mathbb{1})(T^* - \mathrm{i}\mathbb{1})\varphi = 0. \tag{11.8}$$

Put $\eta = (T^* - \mathrm{i}\mathbb{1})\varphi$, and note that it follows from (11.8) that $\eta \in \mathscr{D}_-$. Now for any vector $\eta' \in \mathscr{D}_-$ we have

$$\mathrm{i} \langle \eta' | \eta \rangle = \mathrm{i} \langle \eta' | (T^* - \mathrm{i}\mathbb{1})\varphi \rangle = \langle -\mathrm{i}\eta' | T^*\varphi \rangle + \mathrm{i} \langle \eta' | - \mathrm{i}\varphi \rangle$$

$$= \langle -\mathrm{i}\eta' | T^*\varphi \rangle + \langle \eta' | \varphi \rangle = \left\langle \begin{bmatrix} \eta' \\ -\mathrm{i}\eta' \end{bmatrix} \middle| \begin{bmatrix} \varphi \\ T^*\varphi \end{bmatrix} \right\rangle = 0,$$

because $\left[\begin{smallmatrix} \eta' \\ -\mathrm{i}\eta' \end{smallmatrix} \right] \in \widetilde{\mathscr{D}}_-$ and we have assumed that $\left[\begin{smallmatrix} \varphi \\ T^*\varphi \end{smallmatrix} \right] \perp \widetilde{\mathscr{D}}_-$. Putting $\eta' = \eta$ we obtain $\eta = 0$, and consequently $T^*\varphi = \mathrm{i}\varphi$. This, in turn, means that $\varphi \in \mathscr{D}_+$, i.e. $\left[\begin{smallmatrix} \varphi \\ T^*\varphi \end{smallmatrix} \right] \in \widetilde{\mathscr{D}}_+$. However $\left[\begin{smallmatrix} \varphi \\ T^*\varphi \end{smallmatrix} \right]$ is also orthogonal to $\widetilde{\mathscr{D}}_+$, so that $\varphi = 0$. □

From Theorem 11.6 we can immediately derive sufficient and necessary conditions for T to be self-adjoint.

Corollary 11.7 *Let T be a closed symmetric operator. Then the following conditions are equivalent:*
(1) *T is self-adjoint,*
(2) $\mathscr{D}_+ = \mathscr{D}_- = \{0\}$,
(3) $n_+ = n_- = 0$,
(4) *the Cayley transform of T is a unitary operator.*

Similarly from Corollaries 11.7 and 11.5 we infer the following:

Corollary 11.8 *Let T be a closed symmetric operator. Then the following conditions are equivalent:*
(1) *T has a self-adjoint extension,*
(2) *there exists a unitary operator $\mathscr{D}_+ \to \mathscr{D}_-$,*
(3) $n_+ = n_-$.

Proof
Self-adjoint extensions of T are in bijection with unitary extensions of $\mathring{c_T}$ and such an extension exists if and only if there is a unitary operator $\mathscr{D}_+ \to \mathscr{D}_-$, i.e. if and only if $n_+ = n_-$. □

Note that Corollary 11.8 not only provides information on existence of self-adjoint extensions of a given symmetric operator T, but also on the number of such extensions. More precisely, there are as many extensions of T as there are unitary operators $\mathscr{D}_+ \to \mathscr{D}_-$, i.e. infinitely many in any non-trivial case. In other words, self-adjoint extensions of T are parametrized by the unitary group $\mathrm{U}(n_+)$.

Remark 11.9 A fully analogous analysis can be performed without the assumption that T is closed. In that case one must manage without the useful tool in the form of the z-transform, but the conclusion is practically the same as the results for closed symmetric operators given in Corollaries 11.7 and 11.8. The main difference lies in the fact that if T is not assumed to be closed, then it might happen that both deficiency indices are zero, but T is still not self-adjoint. In this case the closure of T is self-adjoint and we say that T is *essentially self-adjoint*.

We will end this section with a simple criterion for existence of self-adjoint extensions of a given symmetric operator. Let us begin by recalling that a mapping $K : \mathcal{H} \to \mathcal{H}$ is an *anti-linear operator* if it is additive:

$$K(\xi + \eta) = K(\xi) + K(\eta), \qquad \xi, \eta \in \mathcal{H}$$

and we have

$$K(\alpha\xi) = \overline{\alpha}K(\xi), \qquad \alpha \in \mathbb{C},\ \xi \in \mathcal{H}.$$

As for linear operators, we omit parentheses in the notation of an action of an anti-linear operator on a vector, i.e. we write $K\xi$ instead of $K(\xi)$.

An anti-linear operator $J : \mathcal{H} \to \mathcal{H}$ is called *anti-unitary* if J is surjective and

$$\langle J\xi \,|\, J\eta \rangle = \langle \eta \,|\, \xi \rangle, \qquad \xi, \eta \in \mathcal{H}$$

(it is equivalent to J being an isometry of $\mathcal{H}$ onto $\mathcal{H}$). Finally an anti-unitary operator is called an *involution* if $J^2 = \mathbb{1}$.

Proposition 11.10 *Let T be a symmetric operator and let J be an anti-unitary involution on $\mathcal{H}$. Assume that $J\big(\mathsf{D}(T)\big) \subset \mathsf{D}(T)$ and*

$$TJ\psi = JT\psi, \qquad \psi \in \mathsf{D}(T).$$

Then T has a self adjoint extension.

Proof

We have to show that the deficiency indices $n_\pm$ of T are equal. To that end we will check that J maps $\mathscr{D}_+$ bijectively onto $\mathscr{D}_-$. The operator J is $\mathbb{R}$-linear, so $\dim_{\mathbb{R}} \mathscr{D}_+ = \dim_{\mathbb{R}} \mathscr{D}_-$ and hence $\dim \mathscr{D}_+ = \dim \mathscr{D}_-$.

Let $\xi \in \mathscr{D}_\pm = \ker(T^* \mp \mathrm{i}\mathbb{1})$. Then for any $\psi \in \mathsf{D}(T)$ we have

$$\begin{aligned}
\langle J\xi \,|\, T\psi \rangle &= \left\langle J\xi \,\middle|\, J^2 T\psi \right\rangle = \langle J\xi \,|\, JTJ\psi \rangle = \langle TJ\psi \,|\, \xi \rangle \\
&= \left\langle J\psi \,\middle|\, T^*\xi \right\rangle = \langle J\psi \,|\, \pm\mathrm{i}\xi \rangle = \pm\mathrm{i}\,\langle J\psi \,|\, \xi \rangle \\
&= \pm\mathrm{i}\left\langle J\psi \,\middle|\, J^2\xi \right\rangle = \pm\mathrm{i}\,\langle J\xi \,|\, \psi \rangle = \langle \mp\mathrm{i}J\xi \,|\, \psi \rangle,
\end{aligned}$$

and hence $J\xi \in \mathsf{D}(T^*)$ and $T^*J\xi = \mp\mathrm{i}J\xi$. This shows that $J\mathscr{D}_\pm \subset \mathscr{D}_\mp$ and it follows that

$$\mathscr{D}_- = J^2\mathscr{D}_- = J(J\mathscr{D}_-) \subset J\mathscr{D}_+ \subset \mathscr{D}_-.$$

Consequently $J\mathscr{D}_+ = \mathscr{D}_-$. □

11.3 Krein and Friedrichs Extensions

Let T be a densely defined positive operator. Since for any $\psi \in \mathsf{D}(T)$ we have $\langle \psi \,|\, T\psi \rangle \geq 0$, it follows that

$$\langle \psi \,|\, T\psi \rangle = \langle T\psi \,|\, \psi \rangle, \qquad \psi \in \mathsf{D}(T).$$

Hence, using the polarization formula, we obtain

$$\langle\psi|T\phi\rangle = \langle T\psi|\phi\rangle, \qquad \psi, \phi \in \mathsf{D}(T),$$

so that T is symmetric. This section will be devoted to studying self-adjoint extensions of positive operators. As we already mentioned in ▶ Sect. 11.2, we can, without loss of generality, assume that T is closed.

Now let T be positive and self-adjoint. Using spectral theory (or more precisely, functional calculus) one can show that $\sigma(T) \subset \mathbb{R}_+$. However, this fact can be derived using much more elementary methods (cf. ▶ Sect. 9.1):

(A) $\ker(T + \mathbb{1}) = \{0\}$, because $T\psi = -\psi$ for some $\psi \in \mathsf{D}(T)$ implies

$$0 = \langle\psi|T\psi\rangle = -\|\psi\|^2 \leq 0,$$

i.e. $\psi = 0$.

(B) The range of $T + \mathbb{1}$ is dense in $\mathcal{H}$, as $\xi \perp (T + \mathbb{1})\mathcal{H}$ means that

$$\langle\xi|(T + \mathbb{1})\psi\rangle, \qquad \psi \in \mathsf{D}(T),$$

i.e. $\mathsf{D}(T) \ni \psi \mapsto \langle\xi|T\psi\rangle$ is a continuous linear functional and consequently $\xi \in \mathsf{D}(T^*) = \mathsf{D}(T)$ and $T\xi = T^*\xi = -\xi$. In other words $\xi \in \ker(T + \mathbb{1}) = \{0\}$.

(C) For any $\psi \in \mathsf{D}(T)$ we have

$$\begin{aligned}\|(T + \mathbb{1})\psi\|^2 \\ = \|T\psi\|^2 + \langle\psi|T\psi\rangle + \langle T\psi|\psi\rangle + \|\psi\|^2 \geq \|\psi\|^2.\end{aligned} \tag{11.9}$$

In particular the range of $T + \mathbb{1}$ is closed: if $\big((T + \mathbb{1})\psi_n\big)_{n\in\mathbb{N}}$ is a Cauchy sequence then it follows from (11.9) that also $(\psi_n)_{n\in\mathbb{N}}$ is a Cauchy sequence. Let $\phi = \lim_{n\to\infty} \psi_n$. Then from closedness of $T+\mathbb{1}$ we infer that $\lim_{n\to\infty} (T+\mathbb{1})\psi_n = (T+\mathbb{1})\phi$.

Summing up the above observations we see that $T + \mathbb{1}$ is a bijection $\mathsf{D}(T)$ onto $\mathcal{H}$ and the inverse operator $(T+\mathbb{1})^{-1}$ is continuous (in fact it is a contraction). In particular $-1 \notin \sigma(T)$. Furthermore, for any $\lambda > 0$ the operator $\frac{1}{\lambda}T$ is positive and self-adjoint, so $-1 \notin \sigma\big(\frac{1}{\lambda}T\big)$. It follows that $-\lambda \notin \sigma(T)$. This way we have shown that $\sigma(T) \subset \mathbb{R}_+$.

Lemma 11.11 *Let $\mathcal{K}$ be a closed subspace of the Hilbert space $\mathcal{H}$ and let $a \in \mathsf{B}(\mathcal{K})$, $b \in \mathsf{B}(\mathcal{K}, \mathcal{K}^\perp)$ and $c \in \mathsf{B}(\mathcal{K}^\perp)$. Then*

$$\begin{bmatrix} a & b^* \\ b & c \end{bmatrix} \geq 0 \tag{11.10}$$

in $\mathsf{B}(\mathcal{H})$ *if and only if* $a \geq 0$ *in* $\mathsf{B}(\mathcal{K})$ *and for any* $\varepsilon > 0$ *we have*

$$c \geq b(a + \varepsilon \mathbb{1}_{\mathcal{K}})^{-1} b^* \tag{11.11}$$

in $\mathsf{B}(\mathcal{K}^{\perp})$.

Proof

Computing the expectation value $\left\langle \left[\begin{smallmatrix} \xi \\ 0 \end{smallmatrix}\right] \middle| \left[\begin{smallmatrix} a & b^* \\ b & c \end{smallmatrix}\right] \left[\begin{smallmatrix} \xi \\ 0 \end{smallmatrix}\right] \right\rangle$ we easily check that for (11.10) to hold it is necessary that $a \geq 0$ in $\mathsf{B}(\mathcal{K})$.

Let $p_{\mathcal{K}} \in \mathsf{B}(\mathcal{H})$ be the projection onto $\mathcal{K}$. Then for any $\varepsilon > 0$ the operator $\varepsilon p_{\mathcal{K}}$ is positive and $\varepsilon p_{\mathcal{K}} \xrightarrow[\varepsilon \searrow 0]{} 0$ in strong topology (monotonically decreasing). It is clear that (11.10) is equivalent to

$$\begin{bmatrix} a + \varepsilon \mathbb{1}_{\mathcal{K}} & b^* \\ b & c \end{bmatrix} \geq 0, \qquad \varepsilon > 0. \tag{11.12}$$

Assume (11.12) holds. Then for any $\varepsilon > 0$ we have

$$\left\langle \begin{bmatrix} \xi_1 \\ \xi_2 \end{bmatrix} \middle| \begin{bmatrix} a + \varepsilon \mathbb{1}_{\mathcal{K}} & b^* \\ b & c \end{bmatrix} \begin{bmatrix} \xi_1 \\ \xi_2 \end{bmatrix} \right\rangle \geq 0,$$

for arbitrary $\left[\begin{smallmatrix} \xi_1 \\ \xi_2 \end{smallmatrix}\right] \in \mathcal{K} \oplus \mathcal{K}^{\perp}$. In other words

$$\langle \xi_1 | (a + \varepsilon \mathbb{1}_{\mathcal{K}}) \xi_1 \rangle + \left\langle \xi_1 \middle| b^* \xi_2 \right\rangle + \langle \xi_2 | b \xi_1 \rangle + \langle \xi_2 | c \xi_2 \rangle \geq 0, \qquad \xi_1 \in \mathcal{K},\ \xi_2 \in \mathcal{K}^{\perp}.$$

Setting $\xi_1 = -(a + \varepsilon \mathbb{1}_{\mathcal{K}})^{-1} b^* \xi_2$ we obtain

$$\left\langle \xi_2 \middle| -b(a + \varepsilon \mathbb{1}_{\mathcal{K}})^{-1} b^* \xi_2 \right\rangle + \langle \xi_2 | c \xi_2 \rangle \geq 0, \qquad \xi_2 \in \mathcal{K}^{\perp}$$

which is exactly (11.11).

Now assume that $a \geq 0$ and for any $\varepsilon > 0$ the relation (11.11) holds. We have

$$\begin{bmatrix} a + \varepsilon \mathbb{1}_{\mathcal{K}} & b^* \\ b & c \end{bmatrix} = \begin{bmatrix} a + \varepsilon \mathbb{1}_{\mathcal{K}} & b^* \\ b & b(a + \varepsilon \mathbb{1}_{\mathcal{K}})^{-1} b^* \end{bmatrix} + \begin{bmatrix} 0 & 0 \\ 0 & c - b(a + \varepsilon \mathbb{1}_{\mathcal{K}})^{-1} b^* \end{bmatrix}$$

and the element

$$\begin{bmatrix} 0 & 0 \\ 0 & c - b(a + \varepsilon \mathbb{1}_{\mathcal{K}})^{-1} b^* \end{bmatrix}$$

is obviously positive. Therefore, in order to prove (11.12), it is enough to show that for any $\varepsilon > 0$ we have

$$\begin{bmatrix} a + \varepsilon \mathbb{1}_{\mathcal{K}} & b^* \\ b & b(a + \varepsilon \mathbb{1}_{\mathcal{K}})^{-1} b^* \end{bmatrix} \geq 0.$$

Writing A for $a + \varepsilon \mathbb{1}_{\mathcal{K}}$ we get

$$\begin{aligned} &\left\langle \begin{bmatrix} \xi_1 \\ \xi_2 \end{bmatrix} \middle| \begin{bmatrix} a + \varepsilon \mathbb{1}_{\mathcal{K}} & b^* \\ b & b(a + \varepsilon \mathbb{1}_{\mathcal{K}})^{-1} b^* \end{bmatrix} \begin{bmatrix} \xi_1 \\ \xi_2 \end{bmatrix} \right\rangle \\ &\qquad = \langle \xi_1 | A\xi_1 \rangle + \langle \xi_1 | b^*\xi_2 \rangle + \langle \xi_2 | b\xi_1 \rangle + \left\langle \xi_2 \middle| bA^{-1} b^* \xi_2 \right\rangle. \end{aligned}$$

Assume for the moment that

$$\xi_2 = b\eta + \eta' \tag{11.13}$$

with $\eta' \in \ker b^* = (b\mathcal{K})^{\perp}$. Then

$$\begin{aligned} &\left\langle \begin{bmatrix} \xi_1 \\ \xi_2 \end{bmatrix} \middle| \begin{bmatrix} a + \varepsilon \mathbb{1}_{\mathcal{K}} & b^* \\ b & b(a + \varepsilon \mathbb{1}_{\mathcal{K}})^{-1} b^* \end{bmatrix} \begin{bmatrix} \xi_1 \\ \xi_2 \end{bmatrix} \right\rangle \\ &\qquad = \langle \xi_1 | A\xi_1 \rangle + \langle \xi_1 | b^*b\eta + b^*\eta' \rangle \\ &\qquad\qquad + \langle b\eta + \eta' | b\xi_1 \rangle + \left\langle b\eta + \eta' \middle| bA^{-1} b^* (b\eta + \eta') \right\rangle \\ &\qquad = \langle \xi_1 | A\xi_1 \rangle + \langle \xi_1 | b^*b\eta + b^*\eta' \rangle \\ &\qquad\qquad + \langle b^*b\eta + b^*\eta' | \xi_1 \rangle + \left\langle b^*b\eta + b^*\eta' \middle| A^{-1} (b^*b\eta + b^*\eta') \right\rangle \\ &\qquad = \langle \xi_1 | A\xi_1 \rangle + \langle \xi_1 | b^*b\eta \rangle + \langle b^*b\eta | \xi_1 \rangle + \left\langle b^*b\eta \middle| A^{-1} b^*b\eta \right\rangle \\ &\qquad = \langle \xi_1 | A\xi_1 + b^*b\eta \rangle + \left\langle b^*b\eta \middle| \xi_1 + A^{-1} b^*b\eta \right\rangle \\ &\qquad = \left\langle \xi_1 \middle| A(\xi_1 + A^{-1} b^*b\eta) \right\rangle + \left\langle b^*b\eta \middle| \xi_1 + A^{-1} b^*b\eta \right\rangle \\ &\qquad = \left\langle A\xi_1 \middle| \xi_1 + A^{-1} b^*b\eta \right\rangle + \left\langle b^*b\eta \middle| \xi_1 + A^{-1} b^*b\eta \right\rangle \\ &\qquad = \left\langle A\xi_1 + b^*b\eta \middle| \xi_1 + A^{-1} b^*b\eta \right\rangle \\ &\qquad = \left\langle A(\xi_1 + A^{-1} b^*b\eta) \middle| \xi_1 + A^{-1} b^*b\eta \right\rangle \\ &\qquad = \left\langle \xi_1 + A^{-1} b^*b\eta \middle| A(\xi_1 + A^{-1} b^*b\eta) \right\rangle \geq 0. \end{aligned}$$

Now any vector ξ_2 is a limit of vectors of the form (11.13), because $\mathcal{K}^\perp$ is the direct sum $\mathcal{K}^\perp = \overline{b\mathcal{K}} \oplus \ker b^*$. This way we have shown that

$$\left\langle \begin{bmatrix} \xi_1 \\ \xi_2 \end{bmatrix} \middle| \begin{bmatrix} a + \varepsilon \mathbb{1}_\mathcal{K} & b^* \\ b & b(a + \varepsilon \mathbb{1}_\mathcal{K})^{-1} b^* \end{bmatrix} \begin{bmatrix} \xi_1 \\ \xi_2 \end{bmatrix} \right\rangle \geq 0$$

for all $\left[\begin{smallmatrix} \xi_1 \\ \xi_2 \end{smallmatrix}\right] \in \mathcal{K} \oplus \mathcal{K}^\perp$. □

Corollary 11.12 *Let $\mathcal{H}, \mathcal{K}, a, b, c$ be as in Lemma 11.11. Then*

$$0 \leq \begin{bmatrix} a & b^* \\ b & c \end{bmatrix} \leq \mathbb{1} \tag{11.14}$$

if and only is $0 \leq a \leq \mathbb{1}_\mathcal{K}$ and for any $\varepsilon > 0$ we have

$$b(a + \varepsilon \mathbb{1}_\mathcal{K})^{-1} b^* \leq c \leq \mathbb{1}_{\mathcal{K}^\perp} - b(\mathbb{1}_\mathcal{K} - a + \varepsilon \mathbb{1}_\mathcal{K})^{-1} b^*.$$

Proof

By Lemma 11.11 the first inequality of (11.14) is equivalent to $a \geq 0$ and $b(a+\varepsilon\mathbb{1}_\mathcal{K})^{-1}b^* \leq c$ for any $\varepsilon > 0$.

The second inequality of (11.14) is, in turn equivalent to

$$0 \leq \begin{bmatrix} \mathbb{1}_\mathcal{K} - a & -b^* \\ -b & \mathbb{1}_{\mathcal{K}^\perp} - c \end{bmatrix} \tag{11.15}$$

and applying Lemma 11.11 to this case we find that (11.15) is equivalent to $a \leq \mathbb{1}_\mathcal{K}$ and $\mathbb{1}_{\mathcal{K}^\perp} - c \geq b(\mathbb{1}_\mathcal{K} - a + \varepsilon' \mathbb{1}_\mathcal{K})^{-1} b^*$ for any $\varepsilon' > 0$. The latter condition can be expressed also as $c \leq \mathbb{1}_{\mathcal{K}^\perp} - b(\mathbb{1}_\mathcal{K} - a + \varepsilon' \mathbb{1}_\mathcal{K})^{-1} b^*$ for any $\varepsilon' > 0$. This way we proved that (11.14) is equivalent to the condition that $0 \leq a \leq \mathbb{1}_\mathcal{K}$ and

$$b(a + \varepsilon \mathbb{1}_\mathcal{K})^{-1} b^* \leq c \leq \mathbb{1}_{\mathcal{K}^\perp} - b(\mathbb{1}_\mathcal{K} - a + \varepsilon' \mathbb{1}_\mathcal{K})^{-1} b^*, \qquad \varepsilon, \varepsilon' > 0,$$

which is further equivalent to

$$b(a + \varepsilon \mathbb{1}_\mathcal{K})^{-1} b^* \leq c \leq \mathbb{1}_{\mathcal{K}^\perp} - b(\mathbb{1}_\mathcal{K} - a + \varepsilon \mathbb{1}_\mathcal{K})^{-1} b^*, \qquad \varepsilon > 0.$$

□

Now let T and S be positive and self-adjoint operators. We write $T \geq S$ if the bounded operators $(T + \mathbb{1})^{-1}$ and $(S + \mathbb{1})^{-1}$ satisfy $(T + \mathbb{1})^{-1} \leq (S + \mathbb{1})^{-1}$ (cf. Proposition 3.7)

Theorem 11.13
Let T be a closed positive operator. Then there exist positive self-adjoint operators T_K and T_F such that
(1) $T \subset T_K$ *and* $T \subset T_F$,
(2) *a positive self-adjoint operator* $\widetilde{T}$ *is an extension of* T *if and only if* $T_K \leq \widetilde{T} \leq T_F$.

Proof
Let $\mathcal{K}$ be the range of the operator $T + \mathbb{1}$. The arguments used to prove Statements (A) and (C) preceding Lemma 11.11, show that $\mathcal{K}$ is a closed subspace of $\mathcal{H}$ and $T + \mathbb{1}$ is a bijection of $\mathsf{D}(T)$ onto $\mathcal{K}$ not decreasing the norm. It follows that the linear map

$$(T + \mathbb{1})^{-1} : \mathcal{K} \longrightarrow \mathsf{D}(T)$$

is continuous.

Let $p_{\mathcal{K}}$ and $p_{\mathcal{K}^\perp}$ be the projections $\mathcal{H} \to \mathcal{K}$ and $\mathcal{H} \to \mathcal{K}^\perp$ and define

$$a = p_{\mathcal{K}}(T + \mathbb{1})^{-1} \in \mathsf{B}(\mathcal{K}), \qquad b = p_{\mathcal{K}^\perp}(T + \mathbb{1})^{-1} \in \mathsf{B}(\mathcal{K}, \mathcal{K}^\perp).$$

Take an arbitrary $\zeta \in \mathcal{K}$ and let $(T + \mathbb{1})^{-1}\zeta = \xi$. Writing both sides in the decomposition $\mathcal{H} = \mathcal{K} \oplus \mathcal{K}^\perp$ we have

$$\begin{bmatrix} a\zeta \\ b\zeta \end{bmatrix} = \begin{bmatrix} \xi_1 \\ \xi_2 \end{bmatrix},$$

since in this notation $\zeta = \left[\begin{smallmatrix} \zeta \\ 0 \end{smallmatrix}\right]$, we have

$$\begin{aligned} \langle \zeta \,|\, a\zeta \rangle = \langle \zeta \,|\, \xi_1 \rangle &= \left\langle \begin{bmatrix} \zeta \\ 0 \end{bmatrix} \middle| \begin{bmatrix} \xi_1 \\ 0 \end{bmatrix} \right\rangle = \left\langle \begin{bmatrix} \zeta \\ 0 \end{bmatrix} \middle| \begin{bmatrix} \xi_1 \\ \xi_2 \end{bmatrix} \right\rangle \\ &= \langle \zeta \,|\, \xi \rangle = \langle (T + \mathbb{1})\xi \,|\, \xi \rangle = \langle \xi \,|\, (T + \mathbb{1})\xi \rangle \\ &= \|\xi\|^2 + \langle \xi \,|\, T\xi \rangle \geq \|\xi\|^2 = \|a\zeta\|^2 + \|b\zeta\|^2, \end{aligned}$$

because T is symmetric. In other words, we obtain $a \geq a^*a + b^*b = a^2 + b^*b$ which can also be written as

$$a(\mathbb{1}_{\mathcal{K}} - a) \geq b^*b. \tag{11.16}$$

Now let $\widetilde{T}$ be a positive and self-adjoint extension of T. Then $(T + \mathbb{1}) \subset (\widetilde{T} + \mathbb{1})$, and it easily follows that

$$(\widetilde{T} + \mathbb{1})^{-1}|_{\mathcal{K}} = (T + \mathbb{1})^{-1}.$$

In particular the matrix of the operator $(\widetilde{T}+\mathbb{1})^{-1}$ corresponding to the decomposition $\mathcal{H} = \mathcal{K} \oplus \mathcal{K}^{\perp}$ must be of the form

$$(\widetilde{T}+\mathbb{1})^{-1} = \begin{bmatrix} a & b^* \\ b & c \end{bmatrix} \tag{11.17}$$

(b^* in the upper right corner is a consequence of self-adjointness of $(\widetilde{T}+\mathbb{1})^{-1}$). Moreover, since $(\widetilde{T}+\mathbb{1})^{-1}$ is a positive contraction, we have

$$0 \leq \begin{bmatrix} a & b^* \\ b & c \end{bmatrix} \leq \mathbb{1},$$

so by Corollary 11.12

$$b(a+\varepsilon\mathbb{1}_{\mathcal{K}})^{-1}b^* \leq c \leq \mathbb{1}_{\mathcal{K}^{\perp}} - b(\mathbb{1}_{\mathcal{K}} - a + \varepsilon\mathbb{1}_{\mathcal{K}})^{-1}b^* \tag{11.18}$$

for all $\varepsilon > 0$. In particular, for T to have a positive self-adjoint extension $\widetilde{T}$ it is necessary that

$$b(a+\varepsilon\mathbb{1}_{\mathcal{K}})^{-1}b^* \leq \mathbb{1}_{\mathcal{K}^{\perp}} - b(\mathbb{1}_{\mathcal{K}} - a + \varepsilon\mathbb{1}_{\mathcal{K}})^{-1}b^*, \qquad \varepsilon > 0. \tag{11.19}$$

Fix $\varepsilon > 0$. The inequality in (11.19) is equivalent to

$$b\big((a+\varepsilon\mathbb{1}_{\mathcal{K}})^{-1} + (\mathbb{1}_{\mathcal{K}} - a + \varepsilon\mathbb{1}_{\mathcal{K}})^{-1}\big)b^* \leq \mathbb{1}_{\mathcal{K}^{\perp}}. \tag{11.20}$$

Also, we have $(a+\varepsilon\mathbb{1}_{\mathcal{K}})^{-1} + (\mathbb{1}_{\mathcal{K}} - a + \varepsilon\mathbb{1}_{\mathcal{K}})^{-1} = f(a)$, where $f(t) = \frac{1}{t+\varepsilon} + \frac{1}{1-t+\varepsilon}$ for all $t \in [0,1]$. Moreover, since

$$f(t) = \tfrac{1+2\varepsilon}{(t+\varepsilon)(1-t+\varepsilon)},$$

we can rewrite (11.20) as

$$(1+2\varepsilon)b\big((a+\varepsilon\mathbb{1}_{\mathcal{K}})^{-1}(\mathbb{1}_{\mathcal{K}} - a + \varepsilon\mathbb{1}_{\mathcal{K}})^{-1}\big)b^* \leq \mathbb{1}_{\mathcal{K}^{\perp}}$$

or, putting $d = (a+\varepsilon\mathbb{1}_{\mathcal{K}})^{-\frac{1}{2}}(\mathbb{1}_{\mathcal{K}} - a + \varepsilon\mathbb{1}_{\mathcal{K}})^{-\frac{1}{2}}$, as

$$(1+2\varepsilon)bd^*db^* \leq \mathbb{1}_{\mathcal{K}^{\perp}}. \tag{11.21}$$

To prove that (11.21) holds, we note first that (11.16) is simply $F(a) \geq b^*b$, where $F(t) = t(1-t)$ is a continuous function on $[0,1]$ of norm $\frac{1}{4}$. Therefore $b^*b \leq \frac{1}{4}\mathbb{1}_{\mathcal{K}}$ (see Proposition 3.5) and consequently

$$\mathbb{1}_{\mathcal{K}} \geq 4b^*b \geq 2b^*b. \tag{11.22}$$

Now, using (11.22), we compute

$$\begin{aligned}(a+\varepsilon\mathbb{1}_{\mathcal{K}})(\mathbb{1}_{\mathcal{K}}-a+\varepsilon\mathbb{1}_{\mathcal{K}}) &= a(\mathbb{1}_{\mathcal{K}}-a)+\varepsilon(\mathbb{1}_{\mathcal{K}}-a)+\varepsilon^2\mathbb{1}_{\mathcal{K}}\\ &\geq a(\mathbb{1}_{\mathcal{K}}-a)+(\varepsilon+\varepsilon^2)\mathbb{1}_{\mathcal{K}}\\ &\geq b^*b+(\varepsilon+\varepsilon^2)\mathbb{1}_{\mathcal{K}}\\ &\geq b^*b+(\varepsilon+\varepsilon^2)2b^*b\\ &= (1+2\varepsilon)b^*b+\varepsilon^2 b^*b \geq (1+2\varepsilon)b^*b,\end{aligned}$$

which can then be successively rewritten as $d^{-1}d^{*-1} \geq (1+2\varepsilon)b^*b$ and then as $\mathbb{1}_{\mathcal{K}} \geq (1+2\varepsilon)db^*bd^*$. This, in turn, means that $\left\|(1+2\varepsilon)db^*bd^*\right\| \leq 1$. Thus

$$1 \geq (1+2\varepsilon)\|bd^*\|^2 = (1+2\varepsilon)\|db^*\|^2 = (1+2\varepsilon)\|bd^*db^*\|$$

which implies (11.21), and we have proved (11.19).

Now we notice that as $\varepsilon \searrow 0$, the left hand side of (11.19) is monotonically increasing, while the right hand side is monotonically decreasing. Thus, by Theorem 3.15, the left hand side has a supremum in and he right hand side an infimum in $\mathsf{B}(\mathcal{K}^{\perp})$ which are the limits of either side in strong topology as $\varepsilon \searrow 0$. Define

$$c_F = \sup_{\varepsilon>0} b(a+\varepsilon\mathbb{1}_{\mathcal{K}})^{-1}b^*, \qquad c_K = \inf_{\varepsilon>0} \mathbb{1}_{\mathcal{K}^{\perp}} - b(\mathbb{1}_{\mathcal{K}}-a+\varepsilon\mathbb{1}_{\mathcal{K}})^{-1}b^*.$$

Then $c \in \mathsf{B}(\mathcal{K}^{\perp})$ satisfies (11.18) for all $\varepsilon > 0$ if and only if

$$c_F \leq c \leq c_K. \tag{11.23}$$

We will now show that the mapping $\widetilde{T} \mapsto c$ is an order reversing bijection between the set of positive self-adjoint extensions of T and the set of $c \in \mathsf{B}(\mathcal{K}^{\perp})$ satisfying (11.23). We already know that any positive self-adjoint extension $\widetilde{T}$ of T defines c satisfying (11.23) via (11.17). Furthermore, if $\widetilde{T}$ and $\widetilde{T}'$ are positive self-adjoint extensions of T and c and c' are the corresponding operators on $\mathcal{K}^{\perp}$ then

$$\begin{aligned}\left(\widetilde{T} \geq \widetilde{T}'\right) &\Longleftrightarrow \left((\widetilde{T}+\mathbb{1})^{-1} \leq (\widetilde{T}'+\mathbb{1})^{-1}\right)\\ &\Longleftrightarrow \left(\begin{bmatrix} a & b^* \\ b & c\end{bmatrix} \leq \begin{bmatrix} a & b^* \\ b & c'\end{bmatrix}\right)\\ &\Longleftrightarrow \left(\begin{bmatrix} 0 & 0 \\ 0 & c'-c\end{bmatrix} \geq 0\right) \Longleftrightarrow \left(c' \geq c\right).\end{aligned}$$

Finally, if $c \in \mathsf{B}(\mathcal{K}^\perp)$ satisfies (11.23) then

$$0 \le \begin{bmatrix} a & b^* \\ b & c \end{bmatrix} \le \mathbb{1}$$

(by Corollary 11.12). Also, the range of $\left[\begin{smallmatrix} a & b^* \\ b & c \end{smallmatrix}\right]$ is dense in $\mathcal{H}$, because

$$\begin{aligned}\begin{bmatrix} a & b^* \\ b & c \end{bmatrix}\mathcal{H} &= \begin{bmatrix} a & b^* \\ b & c \end{bmatrix}(\mathcal{K}\oplus\mathcal{K}^\perp) \supset \begin{bmatrix} a & b^* \\ b & c \end{bmatrix}(\mathcal{K}\oplus\{0\}) \\ &= \left\{\begin{bmatrix} a\zeta \\ b\zeta \end{bmatrix} \,\middle|\, \zeta \in \mathcal{K}\right\} = (T+\mathbb{1})^{-1}\mathcal{H} = \mathsf{D}(T).\end{aligned}$$

Hence $\ker\left[\begin{smallmatrix} a & b^* \\ b & c \end{smallmatrix}\right] = \{0\}$, and consequently $\left[\begin{smallmatrix} a & b^* \\ b & c \end{smallmatrix}\right]^{-1}$ is a closed densely defined operator. Moreover, it is self-adjoint and positive. In fact it satisfies

$$\begin{bmatrix} a & b^* \\ b & c \end{bmatrix}^{-1} \ge \mathbb{1}.$$

In particular

$$\widetilde{T} = \begin{bmatrix} a & b^* \\ b & c \end{bmatrix}^{-1} - \mathbb{1}$$

is positive and self-adjoint. Clearly $\widetilde{T}$ is an extension of T.

Let T_K and T_F be the positive self-adjoint extensions of T corresponding to $c = c_K$ and $c = c_F$ respectively:

$$(T_K+\mathbb{1})^{-1} = \begin{bmatrix} a & b^* \\ b & c_K \end{bmatrix}, \qquad (T_F+\mathbb{1})^{-1} = \begin{bmatrix} a & b^* \\ b & c_F \end{bmatrix}$$

Then any positive self-adjoint extension $\widetilde{T}$ of T must satisfy

$$T_K \le \widetilde{T} \le T_F \tag{11.24}$$

and any positive self-adjoint operator $\widetilde{T}$ satisfying (11.24) is an extension of T. □

Theorem 11.13 shows that any positive operator has self-adjoint extensions and that among those which are positive there is a minimal and a maximal one (in the sense of the partial order on positive operators). The extensions T_K and T_F are called respectively the *Krein extension* and the *Friedrichs extension* of the positive operator T.

Notes

The reader will find further information on self-adjoint extensions of symmetric operators in [AkGl, Chapter VII], [Mau, Chapter V], [ReSi$_1$, Chapter VIII], [ReSi$_2$, Chapter X]. The Krein and Friedrichs extensions are usually presented in the context of quadratic forms on Hilbert spaces. This approach is taken e.g. in [Kat, Chapter 6], [ReSi$_2$, Chapter X]. The version of the theory presented above is less popular, but requires fewer preliminary results and can be carried out using exclusively bounded operators.

One-Parameter Groups of Unitary Operators

P. Sołtan, *A Primer on Hilbert Space Operators*, Compact Textbooks in Mathematics,
https://doi.org/10.1007/978-3-319-92061-0_12

One of particularly fruitful applications of the theory of operators on Hilbert spaces is in representation theory of topological groups. In this chapter we will study basic properties of representation theory of the abelian group $\mathbb{R}$.

A family $(u_t)_{t\in\mathbb{R}}$ of operators on $\mathcal{H}$ is called a *strongly continuous one-parameter group of unitary operators* if

(1) for any $t \in \mathbb{R}$ the operator u_t is unitary,
(2) for all $t, s \in \mathbb{R}$ we have $u_{t+s} = u_t u_s$,
(3) for any vector $\psi \in \mathcal{H}$ the map

$$\mathbb{R} \ni t \longmapsto u_t\psi \in \mathcal{H} \tag{12.1}$$

is continuous.

In view of (2) condition (3) is equivalent to the fact that for any $\psi \in \mathcal{H}$ the mapping (12.1) is continuous at $0 \in \mathbb{R}$.

Consider the following example: let H be a self-adjoint operator on $\mathcal{H}$ and put $u_t = \exp(-\mathrm{i}tH)$. It is easy to see that conditions (1) and (2) are satisfied. Condition (3) follows from continuity of Borel functional calculus, because as $t \to 0$, the functions $\lambda \mapsto \exp(-\mathrm{i}t\lambda)$ converge pointwise to the constant function 1 (and are all uniformly bounded). It follows that $(u_t)_{t\in\mathbb{R}}$ is a strongly continuous one-parameter group of unitary operators.

In order to shorten our notation we will from now on write $\mathrm{e}^{-\mathrm{i}tH}$ instead of $\exp(-\mathrm{i}tH)$.

12.1 Stone's Theorem

Theorem 12.1
Let H be a self-adjoint operator on $\mathcal{H}$ and for $t \in \mathbb{R}$ let $u_t = \mathrm{e}^{-\mathrm{i}tH}$. Then
(1) *if $\psi \in \mathsf{D}(H)$ then $\lim\limits_{t\to 0} \frac{\mathrm{i}}{t}(u_t\psi - \psi) = H\psi$,*
(2) *if the limit $\lim\limits_{t\to 0} \frac{\mathrm{i}}{t}(u_t\varphi - \varphi)$ exists then $\varphi \in \mathsf{D}(H)$.*

Proof
Ad (1). Take $\psi \in \mathsf{D}(H)$. Then $\psi = (\mathbb{1} + H^2)^{-\frac{1}{2}}\xi$ for some $\xi \in \mathcal{H}$ and we have

$$\frac{\mathrm{i}}{t}(u_t\psi - \psi) = F_t(H)\xi,$$

where

$$F_t(\lambda) = \frac{\mathrm{i}}{t}\frac{\mathrm{e}^{-\mathrm{i}t\lambda}-1}{\lambda}\lambda(1 + \lambda^2)^{-\frac{1}{2}}.$$

The functions F_t are uniformly bounded and they converge pointwise to ζ (defined at the beginning of ▶ Sect. 10.1), as $t \to 0$. Thus

$$\lim_{t\to 0} \frac{\mathrm{i}}{t}(u_t\psi - \psi) = z_H\xi = H\psi.$$

Ad (2). Let us define an operator $\widetilde{H}$ putting

$$\mathsf{D}(\widetilde{H}) = \left\{\varphi \in \mathcal{H} \,\middle|\, \text{the limit } \lim_{t\to 0} \tfrac{\mathrm{i}}{t}(u_t\varphi - \varphi) \text{ exists}\right\}$$

and for $\varphi \in \mathsf{D}(\widetilde{H})$

$$\widetilde{H}\varphi = \lim_{t\to 0} \frac{\mathrm{i}}{t}(u_t\varphi - \varphi).$$

It is clear that $\widetilde{H}$ is a linear operator. Moreover $H \subset \widetilde{H}$, so $\widetilde{H}$ is densely defined. Furthermore, for $\varphi_1, \varphi_2 \in \mathsf{D}(\widetilde{H})$ the calculation

$$\begin{aligned}
\langle\varphi_1 | \widetilde{H}\varphi_2\rangle &= \lim_{t\to 0}\left\langle\varphi_1 \middle| \tfrac{\mathrm{i}}{t}(u_t\varphi_2 - \varphi_2)\right\rangle \\
&= \lim_{t\to 0}\left(\left\langle\varphi_1 \middle| \tfrac{\mathrm{i}}{t}u_t\varphi_2\right\rangle - \left\langle\varphi_1 \middle| \tfrac{\mathrm{i}}{t}\varphi_2\right\rangle\right) \\
&= \lim_{t\to 0}\left(\left\langle\tfrac{-\mathrm{i}}{t}{u_t}^*\varphi_1 \middle| \varphi_2\right\rangle - \left\langle\tfrac{-\mathrm{i}}{t}\varphi_1 \middle| \varphi_2\right\rangle\right) \\
&= \lim_{t\to 0}\left(\left\langle\tfrac{-\mathrm{i}}{t}u_{-t}\varphi_1 \middle| \varphi_2\right\rangle - \left\langle\tfrac{-\mathrm{i}}{t}\varphi_1 \middle| \varphi_2\right\rangle\right) \\
&= \lim_{t\to 0}\left\langle\tfrac{\mathrm{i}}{-t}(u_{-t}\varphi_1 - \varphi_1) \middle| \varphi_2\right\rangle = \langle\widetilde{H}\varphi_1 | \varphi_2\rangle
\end{aligned}$$

shows that $\widetilde{H}$ is symmetric. By Remark 8.9 we have $\widetilde{H} = H$. □

We can draw the following conclusion from Theorem 12.1: let H be a self-adjoint operator and let $\psi_0 \in \mathsf{D}(H)$. Then the initial value problem

$$\mathrm{i}\tfrac{d\Psi}{dt} = H\Psi, \quad \Psi(0) = \psi_0 \tag{12.2}$$

(with unknown function $\Psi : \mathbb{R} \to \mathcal{H}$) has a global solution. Indeed: defining $\Psi(t) = \mathrm{e}^{-\mathrm{i}tH}\psi_0$ we obtain a continuous function $\Psi : \mathbb{R} \to \mathcal{H}$ such that

- $\Psi(0) = \psi_0$,
- Ψ is differentiable at 0,
- $\mathrm{i}\tfrac{d\Psi}{dt}\big|_{t=0} = H\psi_0$.

Moreover, for any $t \in \mathbb{R}$ we have[1]

$$\begin{aligned} \tfrac{1}{s}\big(\Psi(t+s) - \Psi(t)\big) = \mathrm{e}^{-\mathrm{i}tH}\tfrac{1}{s}\big(\Psi(s) - \Psi(0)\big) \underset{s\to 0}{\longrightarrow} \mathrm{e}^{-\mathrm{i}tH}H\psi_0 \\ = H\mathrm{e}^{-\mathrm{i}tH}\psi_0 \\ = H\Psi(t) \end{aligned}$$

which means that Ψ is differentiable everywhere and is the solution of the initial value problem (12.2).

Furthermore, the solution of the problem (12.2) is unique. Indeed: if $\Phi : \mathbb{R} \to \mathcal{H}$ is a solution then putting $f(t) = \|\Phi(t) - \Psi(t)\|^2$ we obtain a differentiable scalar-valued function such that $f(0) = 0$ and

$$\begin{aligned} \tfrac{d}{dt} f(t) &= \tfrac{d}{dt}\langle \Phi(t) - \Psi(t) | \Phi(t) - \Psi(t)\rangle \\ &= \langle -\mathrm{i}H\Phi(t) - \mathrm{i}H\Psi(t) | \Phi(t) - \Psi(t)\rangle \\ &\quad + \langle \Phi(t) - \Psi(t) | -\mathrm{i}H\Phi(t) - \mathrm{i}H\Psi(t)\rangle = 0. \end{aligned}$$

Thus $\Phi(t) = \Psi(t)$ for all t. A similar reasoning is used in the prof of the next theorem.

Theorem 12.2 (Stone's Theorem)
Let $(u_t)_{t\in\mathbb{R}}$ be a strongly continuous one-parameter group of unitary operators on $\mathcal{H}$. Then there exits a self-adjoint operator H on $\mathcal{H}$ such that $u_t = \mathrm{e}^{-\mathrm{i}tH}$.

[1] Application of (10.9) to $f(\lambda) = \lambda$ and $g(\lambda) = \mathrm{e}^{-\mathrm{i}t\lambda}$ yields $\mathrm{e}^{-\mathrm{i}tH}H \subset H\mathrm{e}^{-\mathrm{i}tH}$. Moreover, using Theorem 12.1(2) one can easily show that for any s we have $\mathrm{e}^{\mathrm{i}sH}\big(\mathsf{D}(H)\big) \subset \mathsf{D}(H)$. It follows that $\mathsf{D}\big(H\mathrm{e}^{-\mathrm{i}tH}\big) = \mathrm{e}^{\mathrm{i}tH}\mathsf{D}(H) = \mathsf{D}(H)$, and consequently $\mathrm{e}^{-\mathrm{i}tH}H = H\mathrm{e}^{-\mathrm{i}tH}$.

Proof

For $\varphi \in \mathcal{H}$ and $f \in C_c^\infty(\mathbb{R})$ (smooth functions with compact support) define

$$\varphi_f = \int_{\mathbb{R}} f(t) u_t \varphi \, dt$$

and let $\mathscr{D} = \operatorname{span}\{\varphi_f \mid \varphi \in \mathcal{H},\ f \in C_c^\infty(\mathbb{R})\}$. Now if $(f_n)_{n\in\mathbb{N}}$ is a sequence of smooth functions of compact support such that for any n

- $\operatorname{supp} f_n \subset \left[-\frac{1}{n}, \frac{1}{n}\right]$,
- $\int_{\mathbb{R}} |f_n(t)| \, dt = 1$

then for any $\varphi \in \mathcal{H}$

$$\begin{aligned} \|\varphi_{f_n} - \varphi\| &= \left\| \int_{\mathbb{R}} f_n(t) u_t \varphi \, dt - \int_{\mathbb{R}} f_n(t) \varphi \, dt \right\| \\ &= \left\| \int_{\mathbb{R}} f_n(t)(u_t \varphi - \varphi) \, dt \right\| \\ &\le \int_{\mathbb{R}} |f_n(t)| \|u_t \varphi - \varphi\| \, dt \le \sup_{|t| \le \frac{1}{n}} \|u_t \varphi - \varphi\| \xrightarrow[n\to\infty]{} 0. \end{aligned}$$

It follows that $\mathscr{D}$ is a dense subspace of $\mathcal{H}$.

Note that if $f \in C_c^\infty(\mathbb{R})$ and we denote by f_s the function $t \mapsto f(t-s)$ then

$$\begin{aligned} u_s \varphi_f = u_s \int_{\mathbb{R}} f(t) u_t \varphi \, dt &= \int_{\mathbb{R}} f(t) u_s u_t \varphi \, dt \\ &= \int_{\mathbb{R}} f(t) u_{s+t} \varphi \, dt = \int_{\mathbb{R}} f(t-s) u_t \varphi \, dt = \varphi_{f_s} \end{aligned}$$

which shows that the operators u_s preserve the subspace $\mathscr{D}$. Furthermore, for $\varphi_f \in \mathscr{D}$ we have

$$\begin{aligned} \tfrac{1}{s}(u_s \varphi_f - \varphi_f) &= \tfrac{1}{s}\left(\int_{\mathbb{R}} f(t-s) u_t \varphi \, dt - \int_{\mathbb{R}} f(t) u_t \varphi \, dt \right) \\ &= \int_{\mathbb{R}} \tfrac{f(t-s) - f(t)}{s} u_t \varphi \, dt \xrightarrow[s\to 0]{} \int_{\mathbb{R}} \left(-f'(t)\right) u_t \varphi \, dt = -\varphi_{f'}, \end{aligned}$$

because $\frac{f_s - f}{s} \xrightarrow[s\to 0]{} -f'$ uniformly.

Define a linear operator H_0 on $\mathcal{H}$ putting $\mathsf{D}(H_0) = \mathscr{D}$ and

$$H_0 \phi = \mathrm{i} \lim_{s\to 0} \tfrac{1}{s}(u_s \phi - \phi), \qquad \phi \in \mathscr{D}.$$

Notice that the range of H_0 is contained in $\mathscr{D}$ and for any $t \in \mathbb{R}$ we have $H_0 u_t = u_t H_0$.

Now in the same way as in the proof of Theorem 12.1(2) we show that the operator H_0 is symmetric: for $\phi, \psi \in \mathscr{D}$ we have

$$\begin{aligned}\langle\psi|H_0\phi\rangle &= \lim_{s\to 0}\tfrac{\mathrm{i}}{s}\langle\psi|u_s\phi-\phi\rangle = \lim_{s\to 0}\tfrac{\mathrm{i}}{s}\langle\psi|(u_s-\mathbb{1})\phi\rangle\\ &= \lim_{s\to 0}\tfrac{-\mathrm{i}}{-s}\langle(u_{-s}-\mathbb{1})\psi|\phi\rangle = -\mathrm{i}\langle\mathrm{i}H_0\psi|\phi\rangle = \langle H_0\psi|\phi\rangle\,.\end{aligned}$$

We will now show that H_0 is essentially self-adjoint, i.e. $\overline{H_0}$ is self-adjoint. For this we need to prove that $\ker(H_0{}^* \pm \mathrm{i}\mathbb{1}) = \{0\}$ (see Corollary 11.7). Take $\eta \in \ker(H_0{}^* \pm \mathrm{i}\mathbb{1})$. Then for any $\phi \in \mathsf{D}(H_0)$ we have

$$\begin{aligned}\tfrac{d}{dt}\langle\eta|u_t\phi\rangle &= \lim_{s\to 0}\left\langle\eta\,\middle|\,\tfrac{1}{s}(u_{t+s}-u_t)\phi\right\rangle\\ &= \lim_{s\to 0}\left\langle\eta\,\middle|\,u_t\tfrac{1}{s}(u_s-\mathbb{1})\phi\right\rangle\\ &= \langle\eta|u_t(-\mathrm{i})H_0\phi\rangle = -\mathrm{i}\langle\eta|H_0u_t\phi\rangle\\ &= -\mathrm{i}\left\langle H_0{}^*\eta\middle|u_t\phi\right\rangle = -\mathrm{i}\langle\mp\mathrm{i}\eta|u_t\phi\rangle = \pm\langle\eta|u_t\psi\rangle\,.\end{aligned} \tag{12.3}$$

This means that the scalar-valued function $g : t \mapsto \langle\eta|u_t\phi\rangle$ satisfies $g' = \pm g$, so that

$$g(t) = g(0)\mathrm{e}^{\pm t}, \qquad t \in \mathbb{R}.$$

On the other hand $|g(t)| \le \|\eta\|\|u_t\phi\| = \|\eta\|\|\phi\|$, so g must be constant and hence equal to 0. In particular

$$\langle\eta|\phi\rangle = g(0) = 0.$$

Since this holds for all $\phi \in \mathscr{D}$ and $\mathscr{D}$ is dense in $\mathcal{H}$, we obtain $\eta = 0$.

Put $H = \overline{H_0}$. Then H is a self-adjoint operator and we can consider the one-parameter group $(\mathrm{e}^{-\mathrm{i}tH})_{t\in\mathbb{R}}$. Take $\phi \in \mathscr{D}$ and let

$$\xi(t) = u_t\phi - \mathrm{e}^{-\mathrm{i}tH}\phi, \qquad t \in \mathbb{R}.$$

Then for any t we have $\xi(t) \in \mathscr{D} \subset \mathsf{D}(H)$ and a computation analogous to (12.3) shows that

$$\tfrac{d}{dt}\xi(t) = -\mathrm{i}H_0u_t\phi + \mathrm{i}H\mathrm{e}^{-\mathrm{i}tH}\phi = -\mathrm{i}H\xi(t), \qquad t \in \mathbb{R}.$$

Consequently

$$\begin{aligned}\tfrac{d}{dt}\|\xi(t)\|^2 = \tfrac{d}{dt}\langle\xi(t)|\xi(t)\rangle &= \left\langle\tfrac{d}{dt}\xi(t)\middle|\xi(t)\right\rangle + \left\langle\xi(t)\middle|\tfrac{d}{dt}\xi(t)\right\rangle\\ &= \langle-\mathrm{i}H\xi(t)|\xi(t)\rangle + \langle\xi(t)|-\mathrm{i}H\xi(t)\rangle = 0\end{aligned}$$

and hence the function $t \mapsto \|\xi(t)\|$ is constant. On the other hand $\xi(0) = 0$, so we get $\xi(t) = 0$ for all t, i.e.

$$u_t\phi = \mathrm{e}^{-\mathrm{i}tH}\phi, \qquad t \in \mathbb{R},\ \phi \in \mathscr{D}.$$

Now the density of $\mathscr{D}$ in $\mathcal{H}$ implies that $u_t = \mathrm{e}^{-\mathrm{i}tH}$ for all $t \in \mathbb{R}$. □

Let $(u_t)_{t\in\mathbb{R}}$ be a strongly continuous one-parameter group of unitary operators on $\mathcal{H}$. The self-adjoint operator H such that $u_t = \mathrm{e}^{-\mathrm{i}tH}$ for all t is obviously unique and we call it the *infinitesimal generator* of the group $(u_t)_{t\in\mathbb{R}}$.

12.2 Trotter Formula

Let H be a self-adjoint operator. Recall that $\mathsf{D}(H)$ is a Hilbert space with the graph norm:

$$\|\psi\|_H = \sqrt{\|\psi\|^2 + \|H\psi\|^2}, \qquad \psi \in \mathsf{D}(H).$$

We also know that for any $\xi \in \mathcal{H}$ the function $\mathbb{R} \ni t \mapsto \mathrm{e}^{-\mathrm{i}tH}\xi \in \mathcal{H}$ is continuous and if $\psi \in \mathsf{D}(H)$ then for any t

$$\mathrm{e}^{-\mathrm{i}tH}\psi \in \mathsf{D}(H)$$

and the function $\mathbb{R} \ni t \mapsto \mathrm{e}^{-\mathrm{i}tH}\psi \in \mathsf{D}(H)$ is continuous for the norm $\|\cdot\|_H$. Indeed: it is enough to show continuity at zero (cf. ▶ Sect. 12.1) and this, in light of the fact that $H\mathrm{e}^{-\mathrm{i}tH} = \mathrm{e}^{-\mathrm{i}tH}H$, follows from the following computation

$$\begin{aligned}\left\|\mathrm{e}^{-\mathrm{i}tH}\psi - \psi\right\|_H^2 &= \left\|\mathrm{e}^{-\mathrm{i}tH}\psi - \psi\right\|^2 + \left\|H\left(\mathrm{e}^{-\mathrm{i}tH}\psi - \psi\right)\right\|^2 \\ &= \left\|\mathrm{e}^{-\mathrm{i}tH}\psi - \psi\right\|^2 + \left\|\mathrm{e}^{-\mathrm{i}tH}H\psi - H\psi\right\|^2 \xrightarrow[t\to 0]{} 0.\end{aligned}$$

Theorem 12.3 (Trotter Formula)
Let H and K be self-adjoint operators and assume that $H + K$ is self-adjoint. Then for any $t \in \mathbb{R}$

$$\left(\mathrm{e}^{-\mathrm{i}\frac{t}{n}H}\mathrm{e}^{-\mathrm{i}\frac{t}{n}K}\right)^n \xrightarrow[n\to\infty]{} \mathrm{e}^{-\mathrm{i}t(H+K)}$$

in strong topology.

Proof

For $t \neq 0$ let

$$F(t) = \tfrac{1}{t}\big(e^{-itH}e^{-itK} - e^{-it(H+K)}\big).$$

Now for any $\xi \in \mathcal{H}$ the function $t \mapsto F(t)\xi$ is continuous on $\mathbb{R}\setminus\{0\}$, because the maps

$$\Psi_1 : t \longmapsto \tfrac{1}{t}e^{-it(H+K)}\xi \quad \text{and} \quad \Psi_2 : t \longmapsto \tfrac{1}{t}e^{-itK}\xi$$

are continuous and we have

$$\tfrac{1}{t}e^{-itH}e^{-itK}\xi = e^{-itH}\Psi_2(t).$$

Therefore for $t, t' \in \mathbb{R}\setminus\{0\}$

$$\begin{aligned}
&\big\|\tfrac{1}{t}e^{-itH}e^{-itK}\xi - \tfrac{1}{t}e^{-it'H}e^{-it'K}\xi\big\| \\
&\quad = \big\|e^{-itH}\Psi_2(t) - e^{-it'H}\Psi_2(t')\big\| \\
&\quad \leq \big\|e^{-itH}\Psi_2(t) - e^{-itH}\Psi_2(t')\big\| + \big\|e^{-itH}\Psi_2(t') - e^{-it'H}\Psi_2(t')\big\| \\
&\quad = \big\|\Psi_2(t) - \Psi_2(t')\big\| + \big\|e^{-i(t-t')H}\Psi_2(t') - \Psi_2(t')\big\| \xrightarrow[t\to t']{} 0.
\end{aligned}$$

It is moreover clear that $\lim_{t\to\pm\infty} F(t)\xi = 0$.

Let us denote the domain $\mathsf{D}(H+K) = \mathsf{D}(H)\cap\mathsf{D}(K)$ by D. We will check that for $\psi \in \mathsf{D}$ we also have

$$\lim_{t\to 0} F(t)\psi = 0.$$

Indeed: we have

$$\begin{aligned}
F(t)\psi &= \tfrac{1}{t}\big(e^{-itH}e^{-itK}\psi - \psi\big) - \tfrac{1}{t}\big(e^{-it(H+K)}\psi - \psi\big) \\
&= e^{-itH}\tfrac{1}{t}\big(e^{-itK}\psi - \psi\big) + \tfrac{1}{t}\big(e^{-itH}\psi - \psi\big) - \tfrac{1}{t}\big(e^{-it(H+K)}\psi - \psi\big)
\end{aligned}$$

and since

$$\lim_{t\to 0}\tfrac{1}{t}\big(e^{-it(H+K)}\psi - \psi\big) = -i(H+K)\psi,$$

$$\lim_{t\to 0}\tfrac{1}{t}\big(e^{-itH}\psi - \psi\big) = -iH\psi,$$

$$\lim_{t\to 0}\tfrac{1}{t}\big(e^{-itK}\psi - \psi\big) = -iK\psi,$$

we find that

$$\lim_{t\to 0} F(t)\psi = -\mathrm{i}H\psi - \mathrm{i}K\psi - \big(-\mathrm{i}(H+K)\big)\psi = 0.$$

In particular for $\psi \in \mathsf{D}$ w can put $F(0)\psi = 0$ and this way we obtain a family $\big(F(t)\big)_{t\in\mathbb{R}}$ of linear maps $\mathsf{D} \to \mathcal{H}$ such that the function $t \mapsto F(t)\psi$ is continuous. Moreover

$$\lim_{t\to 0} F(t)\psi = \lim_{t\to\pm\infty} F(t)\psi = 0. \tag{12.4}$$

Consider on D the norm $\|\cdot\|_{H+K}$ in which this space is a Hilbert space. Since $\|\cdot\|_{H+K} \geq \|\cdot\|$, we see that all operators $F(t)$ are continuous from D (with the norm $\|\cdot\|_{H+K}$) to $\mathcal{H}$. It follows from continuity of $t \mapsto F(t)\psi$ and (12.4) that for any $\psi \in \mathsf{D}$ the set

$$\big\{F(t)\psi \,\big|\, t \in \mathbb{R}\big\}$$

is bounded. Therefore, by the Banach-Steinhaus theorem, there exists a constant M such that $\big\|F(t)\big\| \leq M$ for all $t \in \mathbb{R}$, where by $\big\|F(t)\big\|$ we mean the norm of an operator $\big(\mathsf{D}, \|\cdot\|_{H+K}\big) \to \mathcal{H}$. In other words, for any $\psi \in \mathsf{D}$ and $t \in \mathbb{R}$ we have

$$\big\|F(t)\psi\big\| \leq M\|\psi\|_{H+K}.$$

Fix a $\psi \in \mathsf{D}$ and let

$$C = \big\{\mathrm{e}^{-\mathrm{i}s(H+K)}\psi \,\big|\, s \in [-1,1]\big\}.$$

As we explained before formulating the theorem, the function

$$s \longmapsto \mathrm{e}^{-\mathrm{i}s(H+K)}\psi$$

is continuous $\mathbb{R} \to \big(\mathsf{D}, \|\cdot\|_{H+K}\big)$ and hence C is compact in $\big(\mathsf{D}, \|\cdot\|_{H+K}\big)$.

Take now $\varepsilon > 0$ and let $\phi_1, \ldots, \phi_N$ be an $\frac{\varepsilon}{2M}$-net in C. For any $\eta \in C$ there exists i such that $\|\eta - \phi_i\|_{H+K} < \frac{\varepsilon}{2M}$. Furthermore, there exists $\delta > 0$ such that for $|t| < \delta$ we have

$$\big\|F(t)\phi_j\big\| < \tfrac{\varepsilon}{2}, \qquad j = 1, \ldots, N,$$

and therefore

$$\big\|F(t)\eta\big\| \leq \big\|F(t)\eta - F(t)\phi_i\big\| + \big\|F(t)\phi_i\big\| \leq M\|\eta - \phi_i\|_{H+K} + \big\|F(t)\phi_i\big\| < \varepsilon$$

which shows that the functions $\big(\psi \mapsto F(t)\psi\big)_{t\in[-1,1]}$ converge to zero uniformly on C, as $t \to 0$.

In other words, for a fixed $\psi \in \mathsf{D}$ the quantity

$$\big\|F(t)\mathrm{e}^{-\mathrm{i}s(H+K)}\psi\big\|$$

tends to 0, as $t \to 0$, uniformly in $s \in [-1, 1]$.

Now the identity

$$\begin{aligned}
&\left(e^{-i\frac{t}{n}H} e^{-i\frac{t}{n}K}\right)^n \psi - e^{-it(H+K)}\psi \\
&\quad = \left(\left(e^{-i\frac{t}{n}H} e^{-i\frac{t}{n}K}\right)^n - \left(e^{-i\frac{t}{n}(H+K)}\right)^n\right)\psi \\
&\quad = \sum_{m=0}^{n-1} \left(e^{-i\frac{t}{n}H} e^{-i\frac{t}{n}K}\right)^n \left(e^{-i\frac{t}{n}H} e^{-i\frac{t}{n}K} - e^{-i\frac{t}{n}(H+K)}\right)\left(e^{-i\frac{t}{n}(H+K)}\right)^{n-1-m}\psi
\end{aligned}$$

implies that

$$\begin{aligned}
&\left\| \left(e^{-i\frac{t}{n}H} e^{-i\frac{t}{n}K}\right)^n \psi - e^{-it(H+K)}\psi \right\| \\
&\quad \le n \max_{0 \le m \le n-1} \left\| \left(e^{-i\frac{t}{n}H} e^{-i\frac{t}{n}K} - e^{-i\frac{t}{n}(H+K)}\right) e^{-i\frac{t(n-1-m)}{n}(H+K)}\psi \right\| \\
&\quad = |t| \max_{0 \le m \le n-1} \left\| F\left(\tfrac{t}{n}\right) e^{-i\frac{t(n-1-m)}{n}(H+K)}\psi \right\| \\
&\quad \le |t| \max_{|s|<|t|} \left\| F\left(\tfrac{t}{n}\right) e^{-is(H+K)}\psi \right\| \xrightarrow[n\to\infty]{} 0
\end{aligned}$$

which shows that

$$\left(e^{-i\frac{t}{n}H} e^{-i\frac{t}{n}K}\right)^n \psi \xrightarrow[n\to\infty]{} e^{-it(H+K)}\psi \tag{12.5}$$

for $\psi \in \mathsf{D}$. Since $\left\| \left(e^{-i\frac{t}{n}H} e^{-i\frac{t}{n}K}\right)^n \psi \right\| = 1$ for all n, it follows that we have (12.5) for all $\psi \in \mathcal{H}$. □

Notes

Stone's theorem and Trotter formula are standard tools of the theory of operators on Hilbert spaces often applied e.g. in mathematical physics, but also in theory of group representation and many other branches of mathematics. These applications and their far reaching generalizations can be found, among others, in [Kat, Chapter 9], [Mau, Chapter X], [ReSi$_1$, Chapter VIII], [ReSi$_2$, Chapter X].

Appendix A

P. Sołtan, *A Primer on Hilbert Space Operators*, Compact Textbooks in Mathematics,
https://doi.org/10.1007/978-3-319-92061-0

A.1 Banach-Steinhaus Theorem

Theorem A.1
Let X be a Banach space and Y a normed space. Let $\mathcal{T} \subset \mathsf{B}(\mathsf{X}, \mathsf{Y})$ be a family of operators such that for each $x \in \mathsf{X}$ the set $\{Tx \,|\, T \in \mathcal{T}\}$ is bounded. Then there exists a constant $M > 0$ such that for any $T \in \mathcal{T}$ we have $\|T\| \leq M$.

Proof
For $x \in \mathsf{X}$ let us denote

$$M(x) = \sup\left\{\|Tx\| \,\middle|\, T \in \mathcal{T}\right\}.$$

Suppose that $\sup\left\{\|T\| \,\middle|\, T \in \mathcal{T}\right\} = +\infty$. Then there exists a sequence $(T_n)_{n\in\mathbb{N}}$ of elements of $\mathcal{T}$ and a sequence $(y_n)_{n\in\mathbb{N}}$ of unit vectors in X such that

$$\|T_n y_n\| > 4^n, \qquad n \in \mathbb{N}.$$

Putting $x_n = \frac{1}{2^n} y_n$ we obtain a sequence $(x_n)_{n\in\mathbb{N}}$ converging to zero and such that for all n we have $\|T_n x_n\| > 2^n$.

We will now choose a strictly increasing sequence of natural numbers $(n_k)_{k\in\mathbb{N}}$ so that for all $k \in \mathbb{N}$

$$\begin{cases} \|T_{n_{k+1}} x_{n_{k+1}}\| > 1 + k + \sum\limits_{j=1}^{k} M(x_{n_j}), \\ \|x_{n_{k+1}}\| < \frac{1}{2^{k+1}}\Big(\max\left\{\|T_{n_j}\| \,\middle|\, j = 1, \ldots, k\right\}\Big)^{-1}. \end{cases}$$

We put $n_1 = 1$ and once $n_1, \dots, n_p$ have been chosen, we use the facts that $x_n \xrightarrow[n\to\infty]{} 0$ and $\|T_n x_n\| \xrightarrow[n\to\infty]{} +\infty$ to choose n_{p+1} in such a way that $n_{p+1} > n_p$ and

$$\begin{cases} \|T_{n_{p+1}} x_{n_{p+1}}\| > 1 + p + \sum_{j=1}^{p} M(x_{n_j}), \\ \|x_{n_{p+1}}\| < \frac{1}{2^{p+1}} \Big(\max \big\{ \|T_{n_j}\| \,\big|\, j = 1, \dots, p \big\} \Big)^{-1}. \end{cases}$$

In particular, since $\|T_n y_n\| > 4^n$, we have $\|T_n\| \geq 4^n$ for all n, and hence

$$\max \big\{ \|T_{n_j}\| \,\big|\, j = 1, \dots, k \big\} > 1.$$

Consequently for any j

$$\|x_{n_j}\| < \frac{1}{2^j} \Big(\max \big\{ \|T_{n_s}\| \,\big|\, s = 1, \dots, j-1 \big\} \Big)^{-1} < \frac{1}{2^j}.$$

It follows that $\sum_{j=1}^{\infty} \|x_{n_j}\| < +\infty$, so that the series $\sum_{j=1}^{\infty} x_{n_j}$ converges to some $x \in \mathsf{X}$.

Now we have

$$\begin{aligned} \|T_{n_{k+1}} x\| &= \Big\| \sum_{j=1}^{\infty} T_{n_{k+1}} x_{n_j} \Big\| = \Big\| \sum_{j=1}^{k} T_{n_{k+1}} x_{n_j} + T_{n_{k+1}} x_{n_{k+1}} + \sum_{j=k+2}^{\infty} T_{n_{k+1}} x_{n_j} \Big\| \\ &= \Big\| T_{n_{k+1}} x_{n_{k+1}} - \Big(-\sum_{j=1}^{k} T_{n_{k+1}} x_{n_j} - \sum_{j=k+2}^{\infty} T_{n_{k+1}} x_{n_j} \Big) \Big\| \\ &\geq \|T_{n_{k+1}} x_{n_{k+1}}\| - \Big\| \sum_{j=1}^{k} T_{n_{k+1}} x_{n_j} + \sum_{j=k+2}^{\infty} T_{n_{k+1}} x_{n_j} \Big\| \\ &> 1 + k + \sum_{j=1}^{k} M(x_j) - \Big\| \sum_{j=1}^{k} T_{n_{k+1}} x_{n_j} + \sum_{j=k+2}^{\infty} T_{n_{k+1}} x_{n_j} \Big\| \\ &\geq 1 + k + \sum_{j=1}^{k} M(x_j) - \Big\| \sum_{j=1}^{k} T_{n_{k+1}} x_{n_j} \Big\| - \Big\| \sum_{j=k+2}^{\infty} T_{n_{k+1}} x_{n_j} \Big\| \\ &\geq 1 + k + \sum_{j=1}^{k} M(x_j) - \sum_{j=1}^{k} \|T_{n_{k+1}} x_{n_j}\| - \sum_{j=k+2}^{\infty} \|T_{n_{k+1}}\| \|x_{n_j}\|. \end{aligned}$$

The last sum on the right hand side is smaller than $\sum_{j=k+2}^{\infty} \frac{1}{2^j} < 1$, because for $j \geq k+2$ we have

$$\|x_{n_j}\| < \frac{1}{2^j} \Big(\max \big\{ \|T_{n_j}\| \,\big|\, j = 1, \dots, j-1 \big\} \Big)^{-1} \leq \frac{1}{2^j} \frac{1}{\|T_{n_{k+1}}\|}.$$

Furthermore the sum $\sum_{j=1}^{k} \|T_{n_{k+1}} x_{n_j}\|$ is majorized by $\sum_{j=1}^{k} M(x_j)$. This shows that $\|T_{n_{k+1}} x\| > k$, which contradicts boundedness of the set $\{Tx \,|\, T \in \mathcal{T}\}$ and consequently shows that $\sup \{\|T\| \,|\, T \in \mathcal{T}\} < +\infty$. □

A.2 Dynkin's Theorem

Let Ω be a set. A family $\mathcal{P}$ of subsets of Ω is called a $\boldsymbol{\pi}$*-system* if it is closed under finite intersections. Furthermore, a family $\mathcal{L}$ of subsets of Ω is a $\boldsymbol{\lambda}$*-system* if the following conditions are satisfied:

1. $\Omega \in \mathcal{L}$,
2. if $\Delta \in \mathcal{L}$ then $\Delta^{\complement} \in \mathcal{L}$,
3. if $(\Delta_n)_{n \in \mathbb{N}}$ is a sequence of pairwise disjoint elements of $\mathcal{L}$ then $\bigcup_{n=1}^{\infty} \Delta_n \in \mathcal{L}$.

Theorem A.2 (Dynkin's Theorem on π- and λ-Systems)
Let $\mathcal{P}$ be a $\boldsymbol{\pi}$-system and $\mathcal{L}$ a $\boldsymbol{\lambda}$-system of subsets of some set Ω. Assume that $\mathcal{P} \subset \mathcal{L}$. Then the $\boldsymbol{\sigma}$-algebra generated by $\mathcal{P}$ is contained in $\mathcal{L}$.

Remark A.3 Before proving Theorem A.2 let us note that a $\boldsymbol{\lambda}$-system is a $\boldsymbol{\sigma}$-algebra if and only if it is also a $\boldsymbol{\pi}$-system. Indeed: on one hand, if $\mathcal{L}$ is a $\boldsymbol{\sigma}$-algebra, then it is a $\boldsymbol{\pi}$-system. On the other hand, if $\mathcal{L}$ is a $\boldsymbol{\lambda}$-system and a $\boldsymbol{\pi}$-system at the same time then for any sequence $(\Delta_n)_{n \in \mathbb{N}}$ of elements of $\mathcal{L}$ the sets

$$\widetilde{\Delta}_n = \Delta_n \setminus \left(\bigcup_{k=1}^{n-1} \Delta_k \right) = \Delta_n \cap (\Delta_1 \cap \cdots \cap \Delta_{n-1})^{\complement}$$

also belong to $\mathcal{L}$. Moreover $(\widetilde{\Delta}_n)_{n \in \mathbb{N}}$ are pairwise disjoint and

$$\bigcup_{n=1}^{\infty} \Delta_n = \bigcup_{n=1}^{\infty} \widetilde{\Delta}_n \in \mathcal{L}.$$

This shows that $\mathcal{L}$ is closed under arbitrary countable unions of its elements, i.e. it is a $\boldsymbol{\sigma}$-algebra.

Proof of Theorem A.2
Denote by $\boldsymbol{\lambda}(\mathcal{P})$ the smallest $\boldsymbol{\lambda}$-system containing the family $\mathcal{P}$ and by $\boldsymbol{\sigma}(\mathcal{P})$ the smallest $\boldsymbol{\sigma}$-algebra containing $\mathcal{P}$. Clearly $\boldsymbol{\lambda}(\mathcal{P}) \subset \boldsymbol{\sigma}(\mathcal{P})$. We will show that

$$\boldsymbol{\lambda}(\mathcal{P}) = \boldsymbol{\sigma}(\mathcal{P}), \tag{A.1}$$

and hence we will have $\boldsymbol{\sigma}(\mathcal{P}) \subset \mathcal{L}$, because $\boldsymbol{\lambda}(\mathcal{P}) \subset \mathcal{L}$.

To prove (A.1) one needs to show that $\boldsymbol{\lambda}(\mathcal{P})$ is a $\boldsymbol{\sigma}$-algebra. In view of Remark A.3 we have see that $\boldsymbol{\lambda}(\mathcal{P})$ is a $\boldsymbol{\pi}$-system (is closed under intersections).

Fix $\Delta \in \boldsymbol{\lambda}(\mathcal{P})$ and let

$$\mathcal{L}_\Delta = \{\Theta \subset \Omega \,|\, \Theta \cap \Delta \in \boldsymbol{\lambda}(\mathcal{P})\}.$$

We will check that $\mathcal{L}_\Delta$ is a $\boldsymbol{\lambda}$-system:

(1) we have $\Omega \cap \Delta = \Delta \in \boldsymbol{\lambda}(\mathcal{P})$, and so $\Omega \in \mathcal{L}_\Delta$,

(2) if $\Theta \in \mathcal{L}_\Delta$ then

$$\begin{aligned}\Theta^{\mathsf{C}} \cap \Delta = \left(\Theta^{\mathsf{C}} \cup \Delta^{\mathsf{C}}\right) \cap \Delta = (\Theta \cap \Delta)^{\mathsf{C}} \cap \Delta \\ = \left((\Theta \cap \Delta) \cup \Delta^{\mathsf{C}}\right)^{\mathsf{C}} \in \boldsymbol{\lambda}(\mathcal{P}),\end{aligned}$$

because $\Theta \cap \Delta$ and Δ^{C} are disjoint elements of $\boldsymbol{\lambda}(\mathcal{P})$,

(3) if $(\Theta_n)_{n\in\mathbb{N}}$ is a sequence of pairwise disjoint elements of $\mathcal{L}_\Delta$ then

$$\left(\bigcup_{n=1}^{\infty} \Theta_n\right) \cap \Delta = \bigcup_{n=1}^{\infty} (\Theta_n \cap \Delta) \in \boldsymbol{\lambda}(\mathcal{P}),$$

as the latter is a countable union of disjoint elements of $\boldsymbol{\lambda}(\mathcal{P})$.

Now since $\mathcal{P}$ is closed under finite intersections, for any $\Gamma \in \mathcal{P}$ we have $\mathcal{P} \subset \mathcal{L}_\Gamma$, and as $\mathcal{L}_\Gamma$ is a $\boldsymbol{\lambda}$-system, we also have $\boldsymbol{\lambda}(\mathcal{P}) \subset \mathcal{L}_\Gamma$. In other words, for $\Delta \in \boldsymbol{\lambda}(\mathcal{P})$ and $\Gamma \in \mathcal{P}$ we have $\Delta \cap \Gamma \in \boldsymbol{\lambda}(\mathcal{P})$.

This, in turn, means that $\mathcal{P} \subset \mathcal{L}_\Delta$ for any $\Delta \in \boldsymbol{\lambda}(\mathcal{P})$ and consequently

$$\boldsymbol{\lambda}(\mathcal{P}) \subset \mathcal{L}_\Delta. \tag{A.2}$$

Now we notice that (A.2) says that $\boldsymbol{\lambda}(\mathcal{P})$ is a family closed under finite intersections. □

A.3 Tensor Product of Hilbert Spaces

Let $\mathcal{H}$ and $\mathcal{K}$ be Hilbert spaces. Their tensor product as vector spaces will be denoted by $\mathcal{H} \otimes_{\text{alg}} \mathcal{K}$. We will prove that there exists a scalar product on $\mathcal{H} \otimes_{\text{alg}} \mathcal{K}$ such that

$$\langle \psi_1 \otimes \varphi_1 | \psi_2 \otimes \varphi_2 \rangle = \langle \psi_1 | \psi_2 \rangle \langle \varphi_1 | \varphi_2 \rangle, \qquad \psi_1, \psi_2 \in \mathcal{H},\ \varphi_1, \varphi_2 \in \mathcal{K}.$$

Fix $\psi_1 \in \mathcal{H}$ and $\varphi_1 \in \mathcal{K}$ and consider the mapping

$$\mathcal{H} \times \mathcal{K} \ni (\psi_2, \varphi_2) \longmapsto \langle \psi_1 | \psi_2 \rangle \langle \varphi_1 | \varphi_2 \rangle.$$

It is a bi-linear functional, and so it determines a unique linear functional Φ_{ψ_1,φ_1} on $\mathcal{H} \otimes_{\text{alg}} \mathcal{K}$. Let $\mathcal{E}$ be the space of anti-linear functionals on $\mathcal{H} \otimes_{\text{alg}} \mathcal{K}$. Then the map

$$\mathcal{H} \times \mathcal{K} \ni (\psi_1, \varphi_1) \longmapsto \overline{\Phi_{\psi_1,\varphi_1}} \in \mathcal{E}$$

is bi-linear and consequently it defines a linear $\Xi : \mathcal{H} \otimes_{\text{alg}} \mathcal{K} \to \mathcal{E}$ such that

$$\begin{aligned}\big(\Xi(\psi_1 \otimes \varphi_1)\big)(\psi_2 \otimes \varphi_2) &= \overline{\Phi_{\psi_1,\varphi_1}(\psi_2 \otimes \varphi_2)} \\ &= \overline{\langle\psi_1|\psi_2\rangle\,\langle\varphi_1|\varphi_2\rangle} = \langle\psi_2|\psi_1\rangle\,\langle\varphi_2|\varphi_1\rangle\,.\end{aligned}$$

Now for $\xi, \eta \in \mathcal{H} \otimes_{\text{alg}} \mathcal{K}$ let us put

$$\langle\xi|\eta\rangle = \overline{\Xi(\xi)}(\eta) = \overline{\Xi(\xi)(\eta)}. \tag{A.3}$$

It is clear that $\langle\cdot|\cdot\rangle$ is a sesquilinear from. Moreover, taking $\xi = \psi_1 \otimes \varphi_1$ and $\eta = \psi_2 \otimes \varphi_2$ we obtain

$$\langle\psi_1 \otimes \varphi_1|\psi_2 \otimes \varphi_2\rangle = \langle\psi_1|\psi_2\rangle\,\langle\varphi_1|\varphi_2\rangle\,, \qquad \psi_1, \psi_2 \in \mathcal{H},\ \varphi_1, \varphi_2 \in \mathcal{K}.$$

Now if

$$\xi = \sum_{i=1}^{N} \xi_i^1 \otimes \xi_i^2, \qquad \eta = \sum_{j=1}^{M} \eta_j^1 \otimes \eta_j^2,$$

then

$$\begin{aligned}\langle\xi|\eta\rangle = \overline{\Xi(\xi)(\eta)} &= \overline{\Xi\Big(\sum_{i=1}^{N} \xi_i^1 \otimes \xi_i^2\Big)\Big(\sum_{j=1}^{M} \eta_j^1 \otimes \eta_j^2\Big)} \\ &= \overline{\sum_{i=1}^{N}\sum_{j=1}^{M} \Xi(\xi_i^1 \otimes \xi_i^2)(\eta_j^1 \otimes \eta_j^2)} \\ &= \overline{\sum_{i=1}^{N}\sum_{j=1}^{M} \big\langle\eta_j^1\big|\xi_i^1\big\rangle\big\langle\eta_j^2\big|\xi_i^2\big\rangle} \\ &= \sum_{i=1}^{N}\sum_{j=1}^{M} \langle\xi_i^1|\eta_j^1\rangle\langle\xi_i^2|\eta_j^2\rangle.\end{aligned}$$

This immediately shows that

$$\langle\eta|\xi\rangle = \overline{\langle\xi|\eta\rangle}, \qquad \xi, \eta \in \mathcal{H} \otimes_{\text{alg}} \mathcal{K}.$$

Furthermore, for

$$\xi = \sum_{i=1}^{N} \xi_i^1 \otimes \xi_i^2$$

we can always assume that the vectors $\xi_1^2, \ldots, \xi_N^2$ are orthonormal (by applying the Gram-Schmidt orthogonalization process), so if $\xi \neq 0$ then

$$\langle \xi \,|\, \xi \rangle = \sum_{i=1}^{N} \|\xi_1^i\|^2 > 0.$$

This means that (A.3) defines a scalar product on $\mathcal{H} \otimes_{\mathrm{alg}} \mathcal{K}$. Moreover, it is clear that the condition

$$\langle \psi_1 \otimes \varphi_1 \,|\, \psi_2 \otimes \varphi_2 \rangle = \langle \psi_1 \,|\, \psi_2 \rangle \langle \varphi_1 \,|\, \varphi_2 \rangle, \qquad \psi_1, \psi_2 \in \mathcal{H}, \ \varphi_1, \varphi_2 \in \mathcal{K}$$

determines this scalar product uniquely.

This way $\mathcal{H} \otimes_{\mathrm{alg}} \mathcal{K}$ becomes a normed space and we will denote by $\mathcal{H} \otimes \mathcal{K}$ its completion. Clearly $\mathcal{H} \otimes \mathcal{K}$ is a Hilbert space and we call it the *tensor product* of the Hilbert spaces $\mathcal{H}$ and $\mathcal{K}$.

Let $\{\psi_i\}_{i \in I}$ and $\{\varphi_j\}_{j \in J}$ be orthonormal bases in $\mathcal{H}$ and $\mathcal{K}$ respectively. Then the system

$$\{\psi_i \otimes \varphi_j\}_{i \in I,\, j \in J}$$

is orthonormal and it spans a dense subspace of $\mathcal{H} \otimes \mathcal{K}$. It follows that it is an orthonormal basis of $\mathcal{H} \otimes \mathcal{K}$.

Let (Ω, μ) and (Λ, ν) be σ-finite measure spaces such that the Hilbert spaces $\mathcal{H} = L_2(\Omega, \mu)$ and $\mathcal{K} = L_2(\Lambda, \nu)$ are separable. Choose orthonormal bases $\{\psi_i\}_{i \in \mathbb{N}}$ and $\{\varphi_j\}_{j \in \mathbb{N}}$ of $\mathcal{H}$ and $\mathcal{K}$. Then the system of functions

$$\Omega \times \Lambda \ni (\omega, \lambda) \longmapsto \psi_i(\omega)\varphi_j(\lambda), \qquad i, j \in \mathbb{N} \tag{A.4}$$

is orthonormal in $L_2(\Omega \times \Lambda, \mu \otimes \nu)$. We will show that it forms an orthonormal basis of this space. Take $f \in L_2(\Omega \times \Lambda, \mu \otimes \nu)$ and suppose that

$$\int_{\Omega \times \Lambda} \overline{f(\omega, \lambda)} \psi_i(\omega) \varphi_j(\lambda) \, d\mu(\omega) d\nu(\lambda) = 0$$

for all i and j. Notice that since $|f|^2$ is integrable over $\Omega \times \Lambda$, the function $f(\omega, \cdot)$ belongs to $L_2(\Lambda, \nu)$ for almost all λ. Therefore we can write

$$0 = \int_{\Lambda} \Bigg(\int_{\Omega} \overline{f(\omega, \lambda)} \psi_i(\omega) \, d\mu(\omega) \Bigg) \varphi_j(\lambda) \, d\nu(\lambda) = \int_{\Lambda} \langle f(\cdot, \lambda) \,|\, \psi_i \rangle \, \varphi_j(\lambda) \, d\nu(\lambda).$$

Since this holds for all j, the function $\lambda \mapsto \langle f(\cdot,\lambda)|\psi_i\rangle$ is equal to zero almost everywhere, i.e. on $\Lambda\setminus\Delta_i$ for some Δ_i of measure zero. This means that for $\lambda \in \Lambda\setminus\left(\bigcup_{i=1}^{\infty}\Delta_i\right)$ the function $f(\cdot,\lambda)$ is zero almost everywhere on Ω. Consequently f is equal to zero almost everywhere on $\Omega\times\Lambda$.

Since $\{\psi_i\otimes\varphi_j\}_{i,j\in\mathbb{N}}$ is an orthonormal basis of $\mathcal{H}\otimes\mathcal{K}$ and (A.4) is an orthonormal basis of $L_2(\Omega\times\Lambda,\mu\otimes\nu)$, we can define an operator

$$u : L_2(\Omega,\mu)\otimes L_2(\Lambda,\nu) \longrightarrow L_2(\Omega\times\Lambda,\mu\otimes\nu)$$

mapping each $\psi_i\otimes\varphi_j$ to the function

$$\Omega\times\Lambda \ni (\omega,\lambda) \longmapsto \psi_i(\omega)\varphi_j(\lambda) \in \mathbb{C}.$$

As u maps an orthonormal basis onto an orthonormal basis, it is unitary. This proves the following:

Proposition A.4 *Let (Ω,μ) and (Λ,ν) be σ-finite measure spaces such that the Hilbert spaces $\mathcal{H}=L_2(\Omega,\mu)$ and $\mathcal{K}=L_2(\Lambda,\nu)$ are separable. Then the operator*

$$L_2(\Omega,\mu)\otimes L_2(\Lambda,\nu) \longrightarrow L_2(\Omega\times\Lambda,\mu\otimes\nu)$$

mapping $\psi\otimes\varphi$ to the function

$$\Omega\times\Lambda \ni (\omega,\lambda) \longmapsto \psi(\omega)\varphi(\lambda) \in \mathbb{C}$$

is unitary.

A.4 Open Mapping Theorem

In the formulation and proofs of all statements below we will use the convention of Banach space theory according to which the closed unit ball in a Banach space X is denoted by X_1. Also the closed ball with center 0 and radius r in X will be denoted by X_r.

Theorem A.5 (Open Mapping Theorem)
Let X and Y be Banach spaces and let $T\in\mathsf{B}(\mathsf{X},\mathsf{Y})$ be surjective. Then T is an open mapping.

Proof

We will show that $T(\mathsf{X}_1)$ contains a neighborhood of $0 \in \mathsf{Y}$. For any $n \in \mathbb{N}$ define a new norm $\|\cdot\|_n$ on Y setting

$$\|y\|_n = \inf\left\{\|u\|_{\mathsf{X}} + n\|v\|_{\mathsf{Y}} \,\middle|\, u \in \mathsf{X},\ v \in \mathsf{Y},\ y = v + Tu\right\}.$$

Furthermore let

$$\mathsf{Z} = \left\{f : \mathbb{N} \to \mathsf{Y} \,\middle|\, \sup_{n\in\mathbb{N}} \left\|f(n)\right\|_n < +\infty\right\}.$$

With the norm

$$\|f\|_{\mathsf{Z}} = \sup_{n\in\mathbb{N}} \left\|f(n)\right\|_n, \qquad f \in \mathsf{Z}$$

the vector space Z becomes a Banach space.[1]

Define a sequence $(S_n)_{n\in\mathbb{N}}$ of bounded operators from Y to Z:

$$(S_n y)(m) = \begin{cases} y, & m = n, \\ 0, & m \neq n. \end{cases}$$

We have $\|S_n y\|_{\mathsf{Z}} = \|y\|_n$ for all $y \in \mathsf{Y}$.

Take $y \in \mathsf{Y}$ and $n \in \mathbb{N}$. Since $y = y + T0$, we see that

$$\|y\|_n \leq n\|y\|_{\mathsf{Y}},$$

while by surjectivity of T we can choose $x \in \mathsf{X}$ such that $y = 0 + Tx$, and hence

$$\|y\|_n \leq \|x\|_{\mathsf{X}}.$$

The latter estimate does not depend on n, so for any y the set $\{S_n y \mid n \in \mathbb{N}\}$ is bounded in Z. Thus, by the Banach-Steinhaus theorem, there exists a constant $M > 0$ such that $\|S_n\| \leq M$ for all n.

Now fix $\delta \in \left]0, \frac{1}{M}\right[$. We will show that $\overline{T(\mathsf{X}_1)}$ contains the open ball

$$B_{\mathsf{Y}}(\delta) = \left\{y \in \mathsf{Y} \,\middle|\, \|y\|_{\mathsf{Y}} < \delta\right\}.$$

Let $y \in B_{\mathsf{Y}}(\delta)$. Then for any n we have

$$\|y\|_n = \|S_n y\|_{\mathsf{Z}} \leq M\|y\|_{\mathsf{Y}} < M\delta < 1.$$

[1]However, completeness of Z will not be relevant.

It follows that for any n there are $u_n \in \mathsf{X}$ and $v_n \in \mathsf{Y}$ such that $y = v_n + Tu_n$ and

$$\|u_n\|_{\mathsf{X}} + n\|v_n\|_{\mathsf{Y}} < 1.$$

In particular $\|v_n\|_{\mathsf{Y}} < \frac{1}{n}$, and consequently $Tu_n \xrightarrow[n\to\infty]{} y$. Moreover, for any n we have $\|u_n\|_{\mathsf{X}} < 1$, so $y \in \overline{T(\mathsf{X}_1)}$.

We now proceed to show that it follows from $B_{\mathsf{Y}}(\delta) \subset \overline{T(\mathsf{X}_1)}$ that $B_{\mathsf{Y}}\big(\frac{\delta}{2}\big) \subset T(\mathsf{X}_1)$. Let $y \in B_{\mathsf{Y}}\big(\frac{\delta}{2}\big)$. We know that $y \in \overline{T\big(\mathsf{X}_{\frac{1}{2}}\big)}$, so there exists $x_1 \in \mathsf{X}_{\frac{1}{2}}$ such that $\|y - Tx_1\|_{\mathsf{Y}} < \frac{\delta}{4}$. Furthermore since $y - Tx_1 \in B_{\mathsf{Y}}\big(\frac{\delta}{4}\big)$, there exists $x_2 \in \mathsf{X}_{\frac{1}{4}}$ such that $\big\|(y - Tx_1) - Tx_2\big\|_{\mathsf{Y}} < \frac{\delta}{8}$. Continuing this procedure we obtain a sequence $(x_n)_{n\in\mathbb{N}}$ such that for any n we have $\|x_n\|_{\mathsf{X}} < \frac{1}{2^n}$ and

$$\left\| y - \sum_{k=1}^{n} Tx_k \right\|_{\mathsf{Y}} < \frac{\delta}{2^{n+1}}, \qquad n \in \mathbb{N}.$$

It follows that $y = \sum\limits_{k=1}^{\infty} Tx_k = T\bigg(\sum\limits_{k=1}^{\infty} x_k\bigg)$.

This way we proved that for any $x_0 \in \mathsf{X}$ the image of a closed ball with center x_0 under T contains an open ball with center Tx_0. It now easily follows that T is an open map. □

Corollary A.6 *Let $T \in \mathsf{B}(\mathsf{X}, \mathsf{Y})$ be a bijection. Then the inverse of T is bounded.*

Proof

The operator T is an open map, because it is surjective. It follows that T^{-1} is continuous. □

Corollary A.7 (Closed Graph Theorem) *Let $T : \mathsf{X} \to \mathsf{Y}$ be a linear map such that the graph of T, i.e. the set*

$$\mathsf{G}(T) = \left\{ \begin{bmatrix} x \\ Tx \end{bmatrix} \,\middle|\, x \in \mathsf{X} \right\}$$

is closed in $\mathsf{X} \oplus \mathsf{Y}$. Then T is bounded.

Proof

The closed subspace $\mathsf{G}(T) \subset \mathsf{X} \oplus \mathsf{Y}$ is a Banach space (e.g. with the norm $\big\|\big[\begin{smallmatrix} x \\ Tx \end{smallmatrix}\big]\big\| = \max\big\{\|x\|, \|Tx\|\big\}$, cf. ▶ Sect. 8.1). Let P_{X} and P_{Y} be the maps

$$P_{\mathsf{X}} : \mathsf{G}(T) \ni \begin{bmatrix} x \\ Tx \end{bmatrix} \longmapsto x \in \mathsf{X},$$

$$P_{\mathsf{Y}} : \mathsf{G}(T) \ni \begin{bmatrix} x \\ Tx \end{bmatrix} \longmapsto Tx \in \mathsf{X}.$$

P_{X} and P_{Y} are obviously continuous. Moreover P_{X} is a bijection, and hence an invertible operator: $P_{\mathsf{X}}^{-1} \in \mathsf{B}(\mathsf{X}, \mathsf{G}(T))$. Therefore

$$T = P_{\mathsf{Y}} P_{\mathsf{X}}^{-1}$$

is continuous. □

A.5 Quotient Spaces and Algebras

A.5.1 Quotients of Banach Spaces

Let X be a Banach space and let S be a closed subspace of X. The so called *quotient norm* on the quotient vector space X/S is defined as

$$\|x + \mathsf{S}\|_{\mathsf{X}/\mathsf{S}} = \inf_{u \in \mathsf{S}} \|x + u\|_{\mathsf{X}}, \qquad x \in \mathsf{X}.$$

Let us check that this expression does, in fact, define a norm:

- obviously we have $\|\alpha(x + \mathsf{S})\|_{\mathsf{X}/\mathsf{S}} = |\alpha| \|x + \mathsf{S}\|_{\mathsf{X}/\mathsf{S}}$,
- for $x, y \in \mathsf{X}$

$$\begin{aligned} \|(x + \mathsf{S}) + (y + \mathsf{S})\|_{\mathsf{X}/\mathsf{S}} &= \|x + y + \mathsf{S}\|_{\mathsf{X}/\mathsf{S}} \\ &= \inf_{u \in \mathsf{S}} \|x + y + u\|_{\mathsf{X}} = \inf_{u,v \in \mathsf{S}} \|x + y + u + v\|_{\mathsf{X}} \\ &\leq \inf_{u,v \in \mathsf{S}} \big(\|x + u\|_{\mathsf{X}} + \|y + v\|_{\mathsf{X}}\big) \\ &= \Big(\inf_{u \in \mathsf{S}} \|x + u\|_{\mathsf{X}}\Big) + \Big(\inf_{v \in \mathsf{S}} \|y + v\|_{\mathsf{X}}\Big) \\ &= \|x + \mathsf{S}\|_{\mathsf{X}/\mathsf{S}} + \|y + \mathsf{S}\|_{\mathsf{X}/\mathsf{S}}, \end{aligned}$$

- if $\|x + \mathsf{S}\|_{\mathsf{X}/\mathsf{S}} = 0$ then for any n there exists $u_n \in \mathsf{S}$ such that

$$\|x - (-u_n)\|_{\mathsf{X}} = \|x + u_n\|_{\mathsf{X}} \leq \tfrac{1}{n},$$

so that it follows from closedness of S that $x \in \mathsf{S}$, i.e. $x + \mathsf{S} = 0$.

Furthermore X/S with the norm defined above is a Banach space. To prove this let us first notice that the quotient map

$$q : \mathsf{X} \longrightarrow \mathsf{X}/\mathsf{S}$$

is obviously a contraction. Now let $(x_n + \mathsf{S})_{n \in \mathbb{N}}$ be a Cauchy sequence in X/S. Choose a subsequence $(x_{n_k})_{k \in \mathbb{N}}$ of $(x_n)_{n \in \mathbb{N}}$ such that

$$\|(x_{n_k} + \mathsf{S}) - (x_{n_{k+1}} + \mathsf{S})\|_{\mathsf{X}/\mathsf{S}} < \tfrac{1}{2^k}. \tag{A.5}$$

Then let us define a sequence $(u_k)_{k\in\mathbb{N}}$ of elements of S as follows: we put $u_1 = 0$ and choose u_2 such that

$$\|x_{n_1} - x_{n_2} + u_2\|_{\mathsf{X}} \leq \|x_{n_1} - x_{n_2} + \mathsf{S}\|_{\mathsf{X}/\mathsf{S}} + \tfrac{1}{2}.$$

Next we select u_3 such that

$$\big\|(x_{n_2} + u_2) - (x_{n_3} + u_3)\big\|_{\mathsf{X}} \leq \|x_{n_2} - x_{n_3} + \mathsf{S}\|_{\mathsf{X}/\mathsf{S}} + \tfrac{1}{2^2}$$

etc. The procedure yields $(u_k)_{k\in\mathbb{N}}$ such that

$$\big\|(x_{n_k} + u_k) - (x_{n_{k+1}} + u_{k+1})\big\|_{\mathsf{X}} \leq \|x_{n_k} - x_{n_{k+1}} + \mathsf{S}\|_{\mathsf{X}/\mathsf{S}} + \tfrac{1}{2^k}, \qquad k \in \mathbb{N}.$$

Now thanks to the estimate (A.5), we have

$$\big\|(x_{n_k} + u_k) - (x_{n_{k+1}} + u_{k+1})\big\|_{\mathsf{X}} < \tfrac{1}{2^{k-1}}, \qquad k \in \mathbb{N},$$

so that $(x_{n_k} + u_k)_{k\in\mathbb{N}}$ is a Cauchy sequence in X (as the sequence of norms $\big(\|x_{n_k} + u_k\|_{\mathsf{X}}\big)_{k\in\mathbb{N}}$ is summable). Let x_0 be its limit. Then clearly $x_{n_k} + \mathsf{S} = q(x_{n_k} + u_k) \xrightarrow[k\to\infty]{} q(x_0) = x_0 + \mathsf{S}$, because the map q is continuous. But $(x_n + \mathsf{S})_{n\in\mathbb{N}}$ is a Cauchy sequence, so once we found it to have a convergent subsequence (with limit $x_0 + \mathsf{S}$), it also must converge:

$$x_n + \mathsf{S} \xrightarrow[n\to\infty]{} x_0 + \mathsf{S}.$$

In particular X/S is a Banach space.

Theorem A.8
Let X *and* Y *be Banach spaces and* $T \in \mathsf{B}(\mathsf{X}, \mathsf{Y})$. *Put* $\mathsf{S} = \ker T$. *Then the canonical map* $\widetilde{T} : \mathsf{X}/\mathsf{S} \to \mathsf{Y}$ *making the diagram*

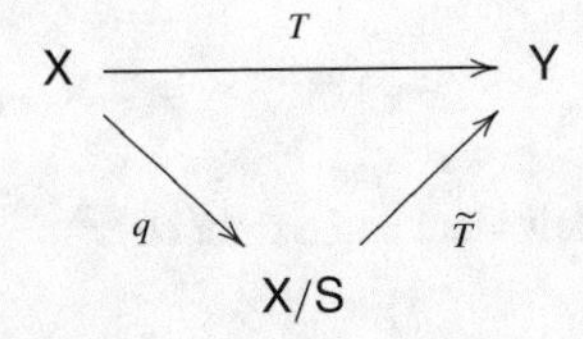

commutative is bounded with $\|\widetilde{T}\| = \|T\|$.

Proof

Take $x \in \mathsf{X}$ and $u \in \mathsf{S}$. Then

$$\|Tx\|_{\mathsf{Y}} = \big\|T(x+u)\big\|_{\mathsf{Y}} \leq \|T\|\|x+u\|_{\mathsf{X}},$$

and hence

$$\|Tx\|_{\mathsf{Y}} \leq \|T\| \inf_{u \in \mathsf{S}} \|x+u\|_{\mathsf{X}} = \|T\|\|x+\mathsf{S}\|_{\mathsf{X}/\mathsf{S}}.$$

Of course $\widetilde{T}(x+\mathsf{S}) = \widetilde{T}(qx) = Tx$, so that $\|\widetilde{T}\| \leq \|T\|$. On the other hand, for any $\varepsilon > 0$ there exists $x_\varepsilon \in \mathsf{X}$ such that $\|x_\varepsilon\|_{\mathsf{X}} = 1$ and $\|Tx\|_{\mathsf{Y}} > \|T\| - \varepsilon$. Thus we have

$$\big\|\widetilde{T}(x_\varepsilon + \mathsf{S})\big\|_{\mathsf{Y}} = \|Tx_\varepsilon\|_{\mathsf{Y}} > \|T\| - \varepsilon,$$

and also $\|x_\varepsilon + \mathsf{S}\|_{\mathsf{X}/\mathsf{S}} \leq \|x_\varepsilon\|_{\mathsf{X}} = 1$. It follows that $\|\widetilde{T}\| \geq \|T\|$. □

Let us note that the closed graph theorem can be used to show that if the range of T in Theorem A.8 is additionally assumed to be closed then $\widetilde{T}$ is an isomorphism of X/S onto $T\mathsf{X}$ (see Corollaries A.6 and A.7).

A.5.2 Ideals and Quotients of C*-Algebras

Let A be a C*-algebra. A *left ideal* in A is a subspace $\mathsf{L} \subset \mathsf{A}$ such that

$$\big(y \in \mathsf{L},\ x \in \mathsf{A}\big) \Longrightarrow \big(xy \in \mathsf{L}\big).$$

Similarly a subspace $\mathsf{R} \subset \mathsf{A}$ is a *right ideal* if

$$\big(y \in \mathsf{R},\ x \in \mathsf{A}\big) \Longrightarrow \big(yx \in \mathsf{R}\big).$$

It is easy to see that L is a left ideal if and only if the set

$$\mathsf{L}^* = \big\{y^* \,\big|\, y \in \mathsf{L}\big\}$$

is a right ideal. An ideal is *two-sided* if it is both a left ideal and a right one.[2]

[2]Generally, the terms "ideal" and "two-sided ideal" are used as synonyms.

Theorem A.9
Let A *be a* C^**-algebra with unit*[3] *and let* $\mathsf{L} \subset \mathsf{A}$ *be a left ideal (not necessarily closed). Then there exists a net* $(e_i)_{i\in I}$ *of elements of* L *such that*
(1) *for any* $i \in I$ *we have* $0 \leq e_i \leq \mathbb{1}$,
(2) $i \preccurlyeq j$ *implies* $e_i \leq e_j$,
(3) *for any* $a \in \mathsf{L}$ *we have* $\|a - ae_i\| \underset{i\in I}{\longrightarrow} 0$.

Proof
Denote by I the set of pairs (n, F) such that $n \in \mathbb{N}$ and F is a finite subset of L. We introduce a partial order on I by declaring that $(n, F) \preccurlyeq (n', F')$ whenever $n \leq n'$ and $F \subset F'$. For $i = (n, F) \in I$ we put

$$v_i = \sum_{b\in F} b^*b, \qquad e_i = \big(\tfrac{1}{n}\mathbb{1} + v_i\big)^{-1} v_i.$$

In other words

$$e_i = f_n(v_i),$$

where $f_n(t) = \frac{t}{\frac{1}{n}+t}$ and since $0 \leq f_n \leq 1$ for all n, we have $0 \leq e_i \leq \mathbb{1}$.
Take $i, j \in I$ such that $i = (n, F) \preccurlyeq j = (n', F')$. Then $v_i \leq v_j$, so that

$$\big(\tfrac{1}{n} + v_i\big)^{-1} \geq \big(\tfrac{1}{n} + v_j\big)^{-1} \tag{A.6}$$

by Proposition 4.25. On the other hand, since for all $t \in \mathbb{R}_+$ we have

$$\tfrac{1}{n}\big(\tfrac{1}{n} + t\big)^{-1} \geq \tfrac{1}{n'}\big(\tfrac{1}{n'} + t\big)^{-1},$$

also the inequality

$$\tfrac{1}{n}\big(\tfrac{1}{n} + v_j\big)^{-1} \geq \tfrac{1}{n'}\big(\tfrac{1}{n'} + v_j\big)^{-1} \tag{A.7}$$

must hold. Putting together (A.6) and (A.7) we obtain

$$\tfrac{1}{n}\big(\tfrac{1}{n} + v_i\big)^{-1} \geq \tfrac{1}{n}\big(\tfrac{1}{n} + v_j\big)^{-1} \geq \tfrac{1}{n'}\big(\tfrac{1}{n'} + v_j\big)^{-1},$$

[3]The assumption of A being unital is not crucial for this theorem (nor for any of the remaining results of this section), but it makes the proof less cumbersome and Statement (1) easier to express.

and hence

$$\mathbb{1} - \tfrac{1}{n}\big(\tfrac{1}{n} + v_i\big)^{-1} \le \mathbb{1} - \tfrac{1}{n'}\big(\tfrac{1}{n'} + v_j\big)^{-1}. \tag{A.8}$$

But

$$e_i = \big(\tfrac{1}{n}\mathbb{1} + v_i\big)^{-1} v_i = \big(\tfrac{1}{n}\mathbb{1} + v_i\big)^{-1}\big(\tfrac{1}{n}\mathbb{1} + v_i - \tfrac{1}{n}\mathbb{1}\big) = \mathbb{1} - \tfrac{1}{n}\big(\tfrac{1}{n} + v_i\big)^{-1} \tag{A.9}$$

and similarly $e_j = \mathbb{1} - \frac{1}{n}\big(\frac{1}{n} + v_j\big)^{-1}$, so (A.8) means that $e_i \le e_j$.

Finally let us take $a \in \mathsf{L}$ and consider $i = (n, F)$ such that $a \in F$. Thanks to (A.9) we have

$$\sum_{b\in F} \big(b(\mathbb{1} - e_i)\big)^*\big(b(\mathbb{1} - e_i)\big) = (\mathbb{1} - e_i)v_i(\mathbb{1} - e_i) = \tfrac{1}{n^2}\big(\tfrac{1}{n} + v_i\big)^{-2} v_i \le \tfrac{1}{4n}\mathbb{1},$$

because $\big(\frac{1}{n} + t\big)^{-2} t \le \frac{n}{4}$ for all $t \in \mathbb{R}_+$. Consequently

$$\big(a(\mathbb{1} - e_i)\big)^*\big(a(\mathbb{1} - e_i)\big) \le \sum_{b\in F} \big(b(\mathbb{1} - e_i)\big)^*\big(b(\mathbb{1} - e_i)\big) \le \tfrac{1}{4n}\mathbb{1},$$

and this gives

$$\|a - ae_i\|^2 = \big\|\big(a(\mathbb{1} - e_i)\big)^*\big(a(\mathbb{1} - e_i)\big)\big\| \le \tfrac{1}{4n}.$$

It follows that $\|a - ae_i\| \xrightarrow[n\to\infty]{} 0$. □

The net $(e_i)_{i\in I}$ constructed in Theorem A.9 is an example of an *approximate unit* for the left ideal L.

Corollary A.10 *Let* J *be a closed ideal in a unital* C^**-algebra* A*. Then* J *is self-adjoint.*

Proof

Let $(e_i)_{i\in I}$ be an approximate unit for J. Then for any $a \in \mathsf{J}$ we have

$$a = \lim_{i\in I} ae_i.$$

Now, since the ideal J is two sided, for any i we have $e_i a^* \in \mathsf{J}$ and consequently

$$a^* = \lim_{i\in I} e_i a^*$$

belongs to the closed subset J. □

Proposition A.11 *Let* J *be a closed ideal in a unital* C^**-algebra* A*. Then the quotient space* A/J *with quotient norm is a* C^**-algebra.*

Proof
Since J is a self-adjoint subset (and a two-sided ideal), it is clear that the quotient space is a unital $*$-algebra. Furthermore, as J is a closed subspace, the quotient space A/J is a Banach space with the quotient norm

$$\|a + \mathsf{J}\| = \inf_{u \in \mathsf{J}} \|a + u\|, \qquad a \in \mathsf{A}.$$

Moreover, the quotient norm satisfies

$$\big\|(a + \mathsf{J})(b + \mathsf{J})\big\| \leq \|a + \mathsf{J}\| \|b + \mathsf{J}\|, \qquad a, b \in \mathsf{A}$$

and

$$\|a^* + \mathsf{J}\| = \|a + \mathsf{J}\|, \qquad a \in \mathsf{A}.$$

The latter of these properties is obvious, while the former is a consequence of the fact that if $\varepsilon > 0$ and $u, v \in \mathsf{J}$ are such that

$$\|a + u\| \leq \|a + \mathsf{J}\| + \varepsilon \quad \text{and} \quad \|b + v\| \leq \|b + \mathsf{J}\| + \varepsilon$$

then

$$\begin{aligned}\|ab + \mathsf{J}\| &\leq \|ab + av + ub + uv\| = \big\|(a + u)(b + v)\big\| \\ &\leq \|a + u\| \|b + v\| \leq \|a + \mathsf{J}\| \|b + \mathsf{J}\| + \varepsilon\big(\|a + \mathsf{J}\| + \|b + \mathsf{J}\| + \varepsilon\big).\end{aligned}$$

It remains to show that for any $a \in \mathsf{A}$ we have the inequality $\|a + \mathsf{J}\|^2 \leq \|a^*a + \mathsf{J}\|$. Let $(e_i)_{i \in I}$ be an approximate unit for J. It is clear that

$$\|a + \mathsf{J}\| \leq \|a - ae_i\|, \qquad i \in I,$$

because $ae_i \in \mathsf{J}$, and hence

$$\|a + \mathsf{J}\| \leq \inf_{i \in I} \|a - ae_i\|.$$

On the other hand, for any $u \in \mathsf{J}$ we have

$$\|a + u\| \geq \big\|(a + u)(\mathbb{1} - e_i)\big\|,$$

since $\|\mathbb{1} - e_i\| \leq 1$. Therefore

$$\begin{aligned}\|a + u\| &\geq \liminf_{i \in I} \big\|(a + u)(\mathbb{1} - e_i)\big\| = \liminf_{i \in I} \big\|(a - ae_i) + (u - ue_i)\big\| \\ &= \liminf_{i \in I} \big\|(a - ae_i)\big\| \geq \inf_{i \in I} \big\|(a - ae_i)\big\|,\end{aligned}$$

where in the first equality we used the fact that $u - ue_i \underset{i\in I}{\longrightarrow} 0$. As a result we get

$$\|a + \mathsf{J}\| = \inf_{i\in I} \|a - ae_i\|, \qquad a \in \mathsf{A}.$$

Using this we compute:

$$\begin{aligned}\|a + \mathsf{J}\|^2 &= \inf_{i\in I} \big\|a(\mathbb{1} - e_i)\big\|^2 = \inf_{i\in I} \big\|(\mathbb{1} - e_i)a^*a(\mathbb{1} - e_i)\big\| \\ &\leq \inf_{i\in I} \big\|a^*a(\mathbb{1} - e_i)\big\| = \|a^*a + \mathsf{J}\|.\end{aligned}$$

□

The first part of the next theorem is a simple generalization of Proposition 4.9 from ▶ Sect. 4.2.

Theorem A.12
Let A *and* B *be unital* C*-*algebras and let*

$$\Phi : \mathsf{A} \longrightarrow \mathsf{B}$$

be a unital $*$*-homomorphism. Then* Φ *is a contraction and* $\Phi(\mathsf{A})$ *is a unital* C*-*subalgebra*[4] *of* B. *If* Φ *is injective then it is isometric.*

Proof
Clearly for any $a \in \mathsf{A}$ the spectrum of the element $\Phi(a)$ in B is contained in the spectrum of a in the algebra A. Now if a is self-adjoint then $\|a\|$ is equal to $|\sigma(a)|$ and $\|\Phi(a)\|$ equals $|\sigma(\Phi(a))|$. Therefore $\|\Phi(a)\| \leq \|a\|$. Now for an arbitrary $b \in \mathsf{A}$, putting $a = b^*b$, we obtain

$$\|\Phi(b)\|^2 = \|\Phi(b^*b)\| = \|\Phi(a)\| \leq \|a\| = \|b^*b\| = \|b\|^2$$

and consequently Φ is a contraction.

Assume now that Φ is injective. In light of the reasoning above, to see that Φ is an isometry, it is enough to prove that it preserves norms of self-adjoint elements. This, in turn, will follow once we establish that for $a = a^*$ we have $\sigma(\Phi(a)) = \sigma(a)$. We already know that $\sigma(\Phi(a)) \subset \sigma(a)$, so suppose that $\sigma(\Phi(a)) \subsetneq \sigma(a)$. Then there exists a continuous function f on $\sigma(a)$ such that $f \neq 0$, but $f = 0$ on $\sigma(\Phi(a))$. We have

$$f(\Phi(a)) = \Phi(f(a)),$$

[4] Closed $*$-subalgebra.

because this holds for polynomials and Φ is continuous. However, since $f = 0$ on $\sigma\big(\Phi(a)\big)$ we find that $f(a) \in \ker\Phi = \{0\}$. In other words $f(a) = 0$ which in view of Theorem 4.23 means that $f = 0$. The resulting contradiction shows that $\sigma\big(\Phi(a)\big) = \sigma(a)$.

Coming back to the case of possibly non-injective Φ we will prove that the range of Φ is closed. Indeed: we can express Φ as the composition of the quotient map $\mathsf{A} \to \mathsf{A}/\ker\Phi$ and the canonical isomorphism $\widetilde{\Phi}$ of $\mathsf{A}/\ker\Phi$ onto $\Phi(\mathsf{A}) \subset \mathsf{B}$. The map $\widetilde{\Phi}$ is an injective unital $*$-homomorphism from the C*-algebra $\mathsf{A}/\ker\Phi$ to the C*-algebra B, so it is isometric and the range of an isometry is closed. □

Remark A.13 An important consequence of the fact that injective $*$-homomorphisms are isometric is the uniqueness of the norm on a C*-algebra (formally speaking we only proved this for unital algebras). This follows by noting that the identity map must then be isometric for any two C*-norms on a given C*-algebra.

Let us consider now a special case, where the arbitrary unital C*-algebra A is replaced by the algebra $\mathrm{C}(X)$ of continuous functions on a compact topological space X. Let J be a closed ideal in $\mathrm{C}(X)$. Furthermore let

$$Y = \big\{x \in X \,\big|\, f(x) = 0 \text{ for all } f \in \mathsf{J}\big\}.$$

Then Y is a closed subset of X, since

$$Y = \bigcap_{f \in \mathsf{J}} f^{-1}(0).$$

Note that the set $X \setminus Y$ is in a natural way a locally compact topological space. Indeed: any compact space is a Tikhonov space, so for any $x_0 \in X \setminus Y$ there exists a continuous function f on X such that $f\big|_Y = 0$ and $f(x_0) = 1$. Then for any compact neighborhood $\mathcal{N}$ of x_0 in X the set

$$\big\{x \in X \,\big|\, f(x) \geq \tfrac{1}{2}\big\} \cap \mathcal{N}$$

is a compact neighborhood of x_0 in $X \setminus Y$.

Denote by $\mathrm{C}_0(X \setminus Y)$ the algebra whose elements are all continuous functions on X vanishing on the subset Y. With norm inherited from $\mathrm{C}(X)$ the algebra $\mathrm{C}_0(X \setminus Y)$ becomes a C*-algebra (without unit). Moreover, it is easy to see that $\mathrm{C}_0(X \setminus Y)$ is naturally isomorphic to the algebra of continuous functions on the topological space $X \setminus Y$ which *vanish at infinity*, i.e. such f that for any $\delta > 0$ the set $\big\{x \in X \setminus Y \,\big|\, |f(x)| \geq \delta\big\}$ is compact in $X \setminus Y$.

Clearly $\mathsf{J} \subset \mathrm{C}_0(X \setminus Y)$. Furthermore

- for any $x \in X \setminus Y$ there exists $f \in \mathsf{J}$ such that $f(x) \neq 0$,
- for any $x_1, x_2 \in X \setminus Y$ there exists $f \in \mathsf{J}$ such that $f(x_1) \neq f(x_2)$ (indeed: there exists $\widetilde{f} \in \mathsf{J}$ such that $\widetilde{f}(x_1) \neq 0$ and there is $g \in \mathrm{C}(X)$ such that $g(x_1) = 1$, $g(x_2) = 0$; it follows that $f = \widetilde{f}g$ satisfies $f \in \mathsf{J}$ and $f(x_1) = g(x_1) \neq 0 = f(x_2)$).

Therefore it follows from the Stone-Weierstrass theorem for locally compact spaces that $\mathsf{J} = \mathrm{C}_0(X \setminus Y)$.

Theorem A.14
Denote $\mathsf{J} = \mathrm{C}_0(X \setminus Y)$. *Then the map*

$$\mathrm{C}(X)/\mathsf{J} \ni f + \mathsf{J} \longmapsto f\big|_Y \in \mathrm{C}(Y)$$

is an isometric $*$*-isomorphism* $\mathrm{C}(X)/\mathrm{C}_0(X \setminus Y)$ *onto* $\mathrm{C}(Y)$.

Proof
The map $\mathrm{C}(X)/\mathsf{J} \ni f + \mathsf{J} \mapsto f\big|_Y$ is a unital $*$-homomorphism which is one-to-one (by definition) and surjective (by Tietze's theorem). Isomorphisms of C*-algebras are isometric. □

Index of Notation

$\mathcal{H}$	Hilbert space
$\mathcal{H}_1$	closed unit ball in $\mathcal{H}$
$\mathcal{H} \oplus \mathcal{K}$	direct sum of Hilbert spaces
$\mathcal{H} \oplus \{0\}$	space of horizontal vectors
$\{0\} \oplus \mathcal{H}$	space of vertical vectors
$\mathcal{H} \otimes_{\mathrm{alg}} \mathcal{K}$	algebraic tensor product
$\mathcal{H} \otimes \mathcal{K}$	tensor product of Hilbert spaces
$\operatorname{span} S$	linear hull of a set S
$\overline{\operatorname{span}} S$	closed linear hull of a set S
$\langle \cdot \mid \cdot \rangle$	scalar product
$\langle \cdot \mid \cdot \rangle_{\mathrm{Tr}}$	scalar product on the space of Hilbert-Schmidt operators
$\langle \psi \mid$	"bra" operator defined by vector ψ
$\mid \psi \rangle$	"ket" operator defined by vector ψ
Tr	trace
$\mathsf{B}(\mathcal{H})$	space of bounded operators on $\mathcal{H}$
$\mathsf{B}(\mathcal{H}, \mathcal{K})$	space of bounded operators $\mathcal{H} \to \mathcal{K}$
$\mathsf{B}(\mathcal{H})_+$	set of positive operators
$\operatorname{Proj}\big(\mathsf{B}(\mathcal{H})\big)$	projections acting on a Hilbert space $\mathcal{H}$
$\mathsf{B}_0(\mathcal{H})$	space of compact operators on $\mathcal{H}$
$\mathsf{B}_1(\mathcal{H})$	space of trace class operators on $\mathcal{H}$
$\mathsf{B}_2(\mathcal{H})$	space of Hilbert-Schmidt operators on $\mathcal{H}$
$\mathcal{F}(\mathcal{H})$	set of finite dimensional operators on $\mathcal{H}$
$\mathbb{1}$	identity operator, unit of a C^*-algebra
$\mathrm{C}(X)$	space of continuous functions on X
$\mathrm{C}_0(X \setminus Y)$	space of continuous functions on $X \setminus Y$ vanishing at infinity
$\mathrm{C}_b(X)$	space of bounded continuous functions on X
$\mathrm{C}_c(X)$	space of continuous functions on X with compact support
$\mathrm{C}_c^\infty(\mathbb{R})$	space of smooth functions on $\mathbb{R}$ with compact support
$\|\cdot\|_\infty$	uniform norm
$\mathscr{B}(X)$	space of bounded Borel functions on X
$\mathbb{C}[\cdot]$	space of polynomials with complex coefficients
X_1 (X_r)	closed unit ball of radius 1 (radius r) in a Banach space X
$\|\cdot\|_{\mathsf{X}/\mathsf{S}}$	quotient norm
$\mathscr{H}(D)$	space of functions holomorphic on a neighborhood of a set D
$\oint$	integral over a closed oriented curve
$L_2(\Omega, \mu)$	space of square-integrable functions on Ω
$L_\infty(\Omega, \mu)$	space of essentially bounded functions on Ω
$\|\cdot\|_2$	norm in L_2 space, Hilbert-Schmidt norm
$\|\cdot\|_1$	trace norm

P. Sołtan, *A Primer on Hilbert Space Operators*, Compact Textbooks in Mathematics,
https://doi.org/10.1007/978-3-319-92061-0

Notation	Meaning
M_f	operator of multiplication by f
$V_{\text{ess}}(f)$	essential range of f
$f \doteq g$	equality almost everywhere
$f \dot{\leq} g$	inequality almost everywhere
χ_Δ	characteristic function of Δ
$\Delta^{\complement}$	complement of Δ
$\mu \otimes \nu$	tensor product of measures (product measure)
$C^*(x)$	C^*-algebra generated by x
$C^*(x, \mathbb{1})$	C^*-algebra generated by x and $\mathbb{1}$
$\sigma(x)$	spectrum of x
$\lvert\sigma(x)\rvert$	spectral radius of x
$\rho(x)$	resolvent set of x
$\sigma(x_1, \ldots, x_n)$	joint spectrum of $x_1, \ldots, x_n$
$\mathbf{l}(x)$	left support of x
$\mathbf{r}(x)$	right support of x
$\mathbf{s}(x)$	support of x
$\operatorname{Re} x, \operatorname{Im} x$	real and imaginary parts of x
$\lvert x \rvert$	modulus of x
$x^{\frac{1}{2}}$	square root of x
x^+, x^-	positive and negative parts of x
$\int_{\sigma(x)} \lambda \, dE_x(\lambda)$	expression of x as a spectral integral
x_f	integral of f with respect to a spectral measure
$\langle \xi \mid E \xi \rangle$	positive measure associated to spectral measure E
$\langle \xi \mid E \eta \rangle$	complex measure associated to spectral measure E
$\mathsf{D}(T)$	domain of T
$\mathsf{G}(T)$	graph of T
$\overline{T}$	closure of T
$\lVert \cdot \rVert_T$	graph norm
T^*	adjoint operator
$\boldsymbol{\zeta}$	the function $t \mapsto \frac{t}{\sqrt{1+t^2}}$
z_T	z-transform of T
$T \subset S$	containment of operators
$\mathring{v}$	isometric operator defined by partial isometry v
$\mathring{c}_T$	Cayley transform of T
$\mathscr{D}_\pm$	deficiency subspaces
$n_\pm$	deficiency indices

References

[AkGl] N.I. Akhiezer, I.M. Glazman, *Theory of Linear Operators in Hilbert Space* (Dover Publications, Mineola, 1993)

[Arv_1] W. Arveson, *An Invitation to* C*-*Algebras* (Springer, New York, 1976)

[Arv_2] W. Arveson, *A Short Course of Spectral Theory* (Springer, New York, 2002)

[Eng] R. Engelking, *General Topology* (Heldermann Verlag, Berlin, 1989)

[Hal] P. Halmos, *A Hilbert Space Problem Book* (Springer, New York, 1982)

[Kat] T. Kato, *Perturbation Theory for Linear Operators* (Springer, Berlin, 1980)

[Lan] E.C. Lance, *Hilbert* C*-*Modules. A Toolkit for Operator Algebraists* (Cambridge University Press, Cambridge, 1995)

[Mau] K. Maurin, *Methods of Hilbert Spaces* (Polish Scientific Publishers, Warsaw, 1972)

[Ped] G.K. Pedersen, *Analysis Now* (Springer, New York, 1995)

[ReSi_1] M. Reed, B. Simon, *Methods of Modern Mathematical Physics I. Functional Analysis* (Academic Press, London, 1980)

[ReSi_2] M. Reed, B. Simon, *Methods of Modern Mathematical Physics II. Fourier Analysis, Self-Adjointness* (Academic Press, London, 1980)

[Rud_1] W. Rudin, *Real and Complex Analysis* (McGraw-Hill, New York, 1987)

[Rud_2] W. Rudin, *Functional Analysis* (McGraw-Hill, New York, 1991)

[WoNa] S.L. Woronowicz, K. Napiórkowski, Operator theory in the C*-algebra framework. Rep. Math. Phys. **31**, 353–371 (1991)

[Zel] W. Żelazko, *Banach Algebras* (Polish Scientific Publishers, Warsaw, 1973)

P. Sołtan, *A Primer on Hilbert Space Operators*, Compact Textbooks in Mathematics,
https://doi.org/10.1007/978-3-319-92061-0

Index

P. Sołtan, *A Primer on Hilbert Space Operators*, Compact Textbooks in Mathematics,
https://doi.org/10.1007/978-3-319-92061-0

Printed in the United States
By Bookmasters